MATHS PLUS

MENTALS AND HOMEWORK BOOK

AUSTRALIAN CURRICULUM

Harry O'Brien
Greg Purcell

OXFORD
UNIVERSITY PRESS

Contents

Contents

Australian Curriculum Linking Chart

Units	1	2	3	4
Number and Algebra				
Interpret, compare and order numbers with more than 2 decimal places, including numbers greater than one, using place value understanding; represent these on a number line (AC9M5N01)				
Express natural numbers as products of their factors, recognise multiples and determine if one number is divisible by another (AC9M5N02)				
Compare and order fractions with the same and related denominators including mixed numerals, applying knowledge of factors and multiples; represent these fractions on a number line (AC9M5N03)				
Recognise that 100% represents the complete whole and use percentages to describe, represent and compare relative size; connect familiar percentages to their decimal and fraction equivalents (AC9M5N04)				
Solve problems involving addition and subtraction of fractions with the same or related denominators, using different strategies (AC9M5N05)				
Solve problems involving multiplication of larger numbers by one- or two-digit numbers, choosing efficient calculation strategies and using digital tools where appropriate; check the reasonableness of answers (AC9M5N06)				
Solve problems involving division, choosing efficient strategies and using digital tools where appropriate; interpret any remainder according to the context and express results as a whole number, decimal or fraction (AC9M5N07)				
Check and explain the reasonableness of solutions to problems including financial contexts using estimation strategies appropriate to the context (AC9M5N08)				
Use mathematical modelling to solve practical problems involving additive and multiplicative situations including financial contexts; formulate the problems, choosing operations and efficient calculation strategies, using digital tools where appropriate; interpret and communicate solutions in terms of the situation (AC9M5N09)				
Create and use algorithms involving a sequence of steps and decisions and digital tools to experiment with factors, multiples and divisibility; identify, interpret and describe emerging patterns (AC9M5N010)				
Recognise and explain the connection between multiplication and division as inverse operations and use this to develop families of number facts (AC9M5A01)				
Find unknown values in numerical equations involving multiplication and division using the properties of numbers and operations (AC9M5A02)				
Measurement and Space				
Choose appropriate metric units when measuring the length, mass and capacity of objects; use smaller units or a combination of units to obtain a more accurate measure (AC9M5M01)				
Solve practical problems involving the perimeter and area of regular and irregular shapes using appropriate metric units (AC9M5M02)				
Compare 12- and 24-hour time systems and solve practical problems involving the conversion between them (AC9M5M03)				
Estimate, construct and measure angles in degrees, using appropriate tools including a protractor, and relate these measures to angle names (AC9M5M04)				
Connect objects to their nets and build objects from their nets using spatial and geometric reasoning (AC9M5SP01)				
Construct a grid coordinate system that uses coordinates to locate positions within a space; use coordinates and directional language to describe position and movement (AC9M5SP02)				
Describe and perform translations, reflections and rotations of shapes, using dynamic geometric software where appropriate; recognise what changes and what remains the same, and identify any symmetries (AC9M5SP03)				
Statistics and Probability				
Acquire, validate and represent data for nominal and ordinal categorical and discrete numerical variables, to address a question of interest or purpose using software including spreadsheets; discuss and report on data distributions in terms of highest frequency (mode) and shape, in the context of the data (AC9M5ST01)				
Interpret line graphs representing change over time; discuss the relationships that are represented and conclusions that can be made (AC9M5ST02)				
Plan and conduct statistical investigations by posing questions or identifying a problem and collecting relevant data; choose appropriate displays and interpret the data; communicate findings within the context of the investigation (AC9M5ST03)				
List the possible outcomes of chance experiments involving equally likely outcomes and compare to those which are not equally likely (AC9M5P01)				
Conduct repeated chance experiments including those with and without equally likely outcomes, observe and record the results; use frequency to compare outcomes and estimate their likelihoods (AC9M5P02)				

Units

5 6 7 8 9 10 11 12 13 14 15 16 17 18 19 20 21 22 23 24 25 26 27 28 29 30 31 32 33 34 35

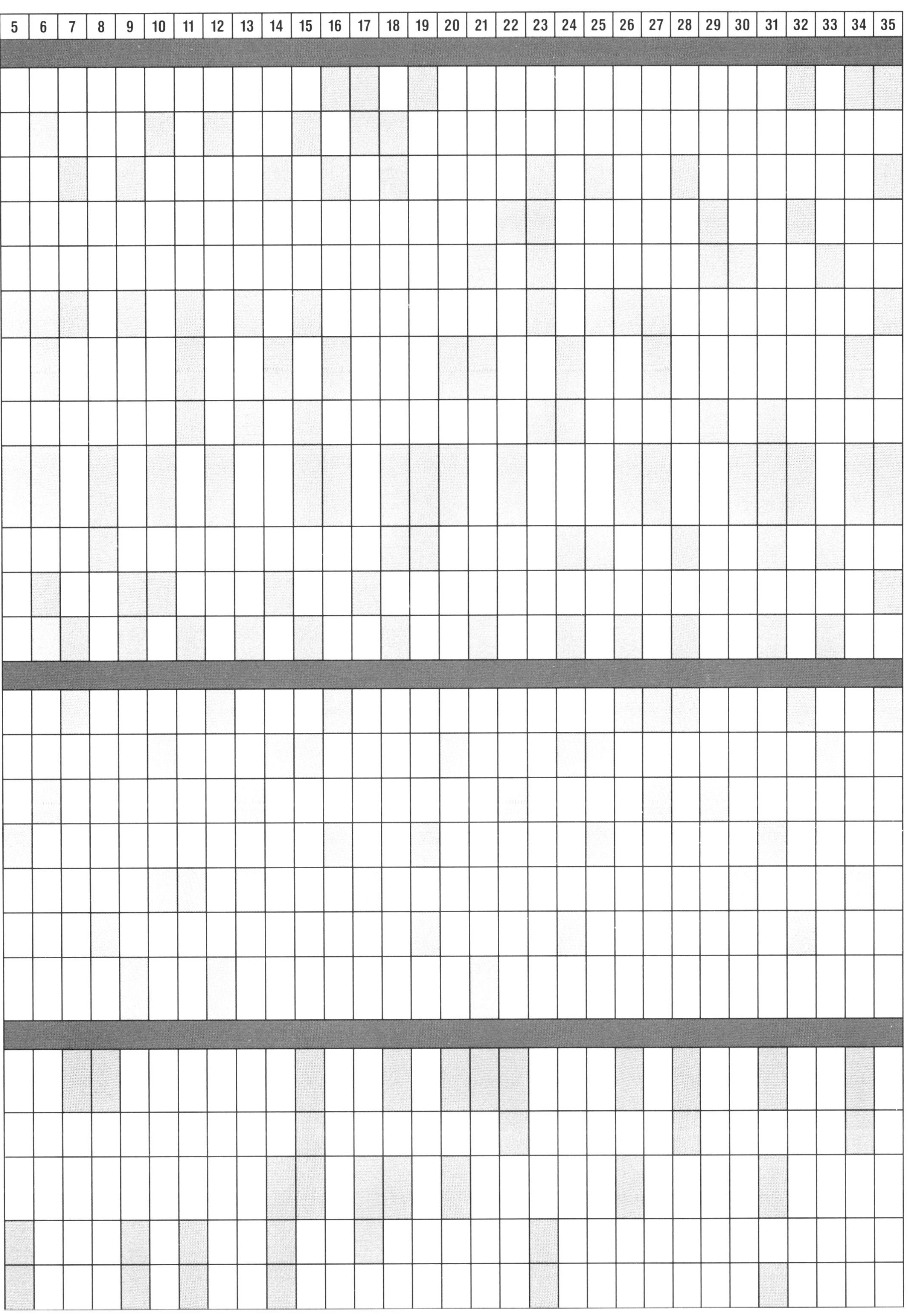

Number and Algebra

SET 1 Basic

1 7 + 5

2 11 – 5

3 8 + 7

4 12 – 6

5 4 × 4

6 5 × 5

7 6 × 6

8 1368c = $ ☐

9 12 ÷ 4

10 20 ÷ 5

11 What is the sum of 16 and 7?

12 What is the product of 5 and 8?

13 Divide 24 by 4.

14 What is the value of 7 in 7326?

15 

Peter had $87 but spent $36. How much did he have left?

$ ☐

SET 2 Addition and subtraction strategies

	236	+	153
Think!	200	+	100
plus	30	+	50
plus	6	+	3
equals	389		

Use the split strategy to answer these questions.

1 254 + 134

2 865 – 343

3 358 + 221

4 789 – 427

5 463 + 435

6 375 + 124

7 423 + 254

8 468 – 253

9 227 + 721

10 677 – 205

Estimate the answer then use the compensation strategy to calculate the exact answer.

	Equation	Estimate	Answer
11	616 + 198		
12	587 – 392		
13	803 + 689		
14	499 + 604		
15	924 – 717		

Space Lines of symmetry

Draw the other half of each shape given its line of symmetry.

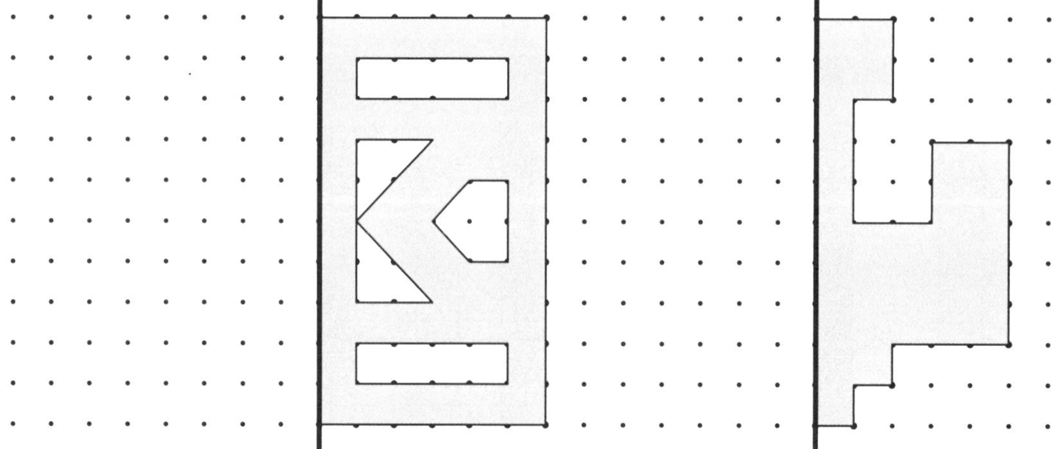

Number and Algebra

SET 3 Multiplication facts

1 What is 6 lots of 3?

2 Multiply three by nine.

3 Add 6 + 6 + 6 + 6.

4 Two lots of 9 plus two more lots

5 Three lots of 6 balls

6 What is 4 lots of 9?

7 What is nine times 6?

8 Multiply 50 by 6.

9 9 + 9 + 9

10 What is 8 lots of 50?

11 Does 8 × 4 equal 4 × 8?

12 Does 7 × 6 equal 6 × 7?

13 What is 10 lots of 9?

14

Name	Rick	Sue	Gillian
Age	32		

Read the clues to complete the table. Gillian is 3 times Sue's age and Sue is $\frac{1}{4}$ of Rick's age.

Mathematical Reasoning

SET 4 Extension

1 3 m = ☐ cm

2 0.37 = ☐ hundredths

3 2000 + 400 + 90 + 6

4 Are 25 and 40 multiples of 5?

5 Share $45 between 5 children.

6 Write the largest number you can using 1, 4, 2, 6, 8.

7 How many faces has a triangular prism?

8 How much are 4 books at $3.50 each?

9 How many minutes between 12:00 and 1:37?

10 Average 11, 13, 17 and 23.

11 What fraction of June is 15 days?

12 Write the set of factors for 28.

13 1500 g of butter at $3.00 per kilogram

14 $\frac{3}{4}$ of $60

15 Write 26 000 in words.

☐

Measurement The kilometre

Colour each box to describe the unit of measurement you would use to measure the item.

	Items	mm	cm	m	km
1	Murray River				
2	thickness of coin				
3	Stuart Highway				
4	height of tree				
5	Olympic marathon				
6	backyard fence				

UNIT 2

Number and Algebra

SET 1 Basic

1 16 – 6

2 24 – 6

3 24 + 6

4 18 + 6

5 4 × 8

6 6 × 4

7 7 × 3

8 $32.15 = ☐ c

9 12 ÷ 6

10 15 ÷ 5

11 What is the sum of 35 and 20?

12 What is the product of 20 and 2?

13 Divide 30 by 3.

14 What is the value of 6 in 6312?

15

Pauline saved $7 a week for 5 weeks. How much did she save?

$ ☐

SET 2 4-digit addition

1

	TH	H	T	O
	3	0	6	4
+	2	7	7	5

2

	TH	H	T	O
	6	5	0	8
+	2	9	5	7

3

	TH	H	T	O
	4	9	0	8
+	3	7	2	5

4

	TH	H	T	O
	8	9	6	5
+	1	0	2	9

5

	TH	H	T	O
	3	0	9	6
+	5	9	6	7

6

	TH	H	T	O
	4	7	1	3
+	4	0	0	9

7 Add 1035, 50 and 200.

8 3000 + 450 + 230

9 What is the total of 170, 230 and 500?

10 800 + 400 + 600

11 What is 2300 more than 3100?

12 What is the sum of 5550 and 1150?

Number and Algebra Multiplication to solve division

Use multiplication and division facts to solve the number sentences.

1 12 ÷ 3 = ☐

2 16 ÷ 2 = ☐

3 15 ÷ 5 = ☐

4 20 ÷ 4 = ☐

5 25 ÷ 5 = ☐

6 24 ÷ 6 = ☐

7 24 ÷ 8 = ☐

8 20 ÷ 5 = ☐

9 30 ÷ 6 = ☐

10 36 ÷ 6 = ☐

11 30 ÷ 5 = ☐

12 40 ÷ 8 = ☐

13 56 ÷ 7 = ☐

14 49 ÷ 7 = ☐

15 72 ÷ 8 = ☐

Number and Algebra

SET 3 Unit fractions

Write a fraction for each shaded shape.

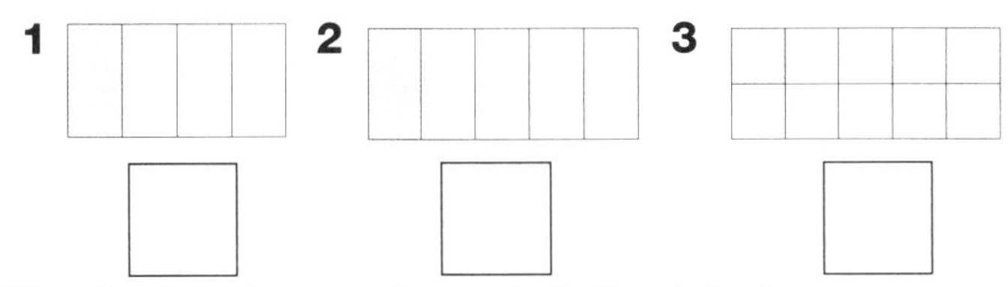

Shade the shapes to match the labels.

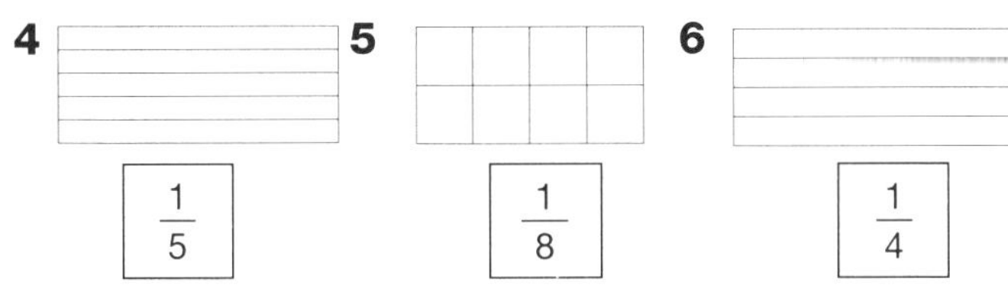

Shade the given fraction of each group.

7 ◯ ◯ ◯ ◯ ◯ ◯ $\frac{1}{6}$

8 △ △ △ $\frac{1}{3}$

9 ◯◯◯◯◯◯◯◯◯◯◯◯ $\frac{1}{12}$

10 ▢ ▢ ▢ ▢ ▢ ▢ $\frac{1}{6}$

Which is larger:

11 $\frac{1}{4}$ or $\frac{1}{5}$?

12 $\frac{1}{5}$ or $\frac{1}{10}$?

13 $\frac{1}{8}$ or $\frac{1}{10}$?

SET 4 Extension

1 $\frac{3}{5}$ of 30

2 How many days in September, April and June?

3 If 3 kg costs $21, how much would 7 kg cost?

4 36 127 + 3041

5 How many $\frac{1}{3}$ s in 2 wholes?

6 How many hours between 9 am and 7 pm?

7 How many edges has a triangular pyramid?

8 232, 257, 282, ▢ , ▢

9 Which is the larger, $\frac{1}{2}$ or $\frac{1}{10}$?

10 Write the smallest number you can using 9, 1, 7, 6, 4.

11 Share 96 between 6 people.

12 How much is $4\frac{1}{2}$ kg of chicken at $4.80 per kilogram.

13 Write the set of factors for 32.

14 $2\frac{1}{4}$ years = ▢ months.

15 Write 39 006 in words.

Measurement Calculating area

Record the length and width of these shapes and then calculate the area.

	Length	Width	Area (L × W)
1			
2			
3			
4			

1

2 cm

4 cm

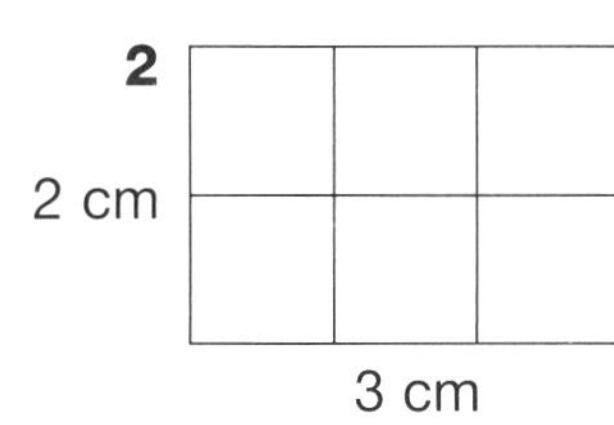

3

4 cm

3 cm

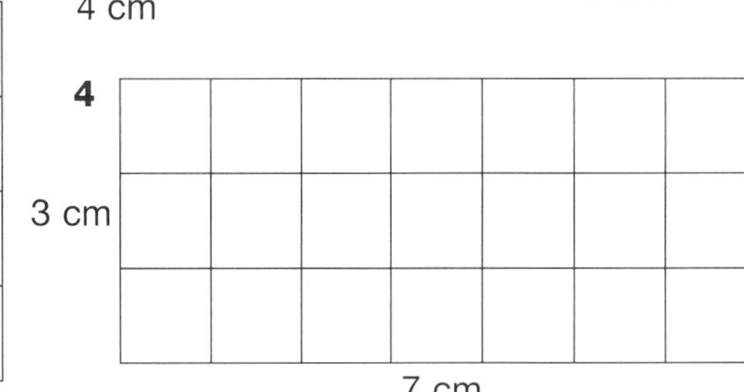

Number and Algebra

SET 1 Basic

1 6 + 5

2 9 + 4

3 15 – 4

4 5 × 8

5 8 – 7

6 9 × 6

7 7 × 2

8 $4.21 = ☐ c

9 18 ÷ 3

10 21 ÷ 3

11 What is the product of 6 and 2?

12 What is the sum of 8 and 4?

13 What is the value of 3 in 5382?

14 Divide 21 by 7.

15 

Shaun shared 63 marbles among himself and 6 friends. How many did each child receive?

☐ marbles

SET 2 Subtracting 4-digit numbers

1

	TH	H	T	O
	5	8	3	4
–	4	2	7	3

2

	TH	H	T	O
	6	8	4	5
–	3	2	7	7

3

	TH	H	T	O
	2	5	0	8
–		4	2	8

4

	TH	H	T	O
	5	6	6	0
–		2	4	5

5 Subtract 137 from 899.

6 767 take away 347

7 907 minus 603

8 $9.75 – $3.30

9 Mr Brown, the baker, baked 1050 cakes for the fete but sold only 742. How many were left? 

10 Jim needed to save $598 to buy a surfboard. If he saved $246, how much more did he need to save? 

Space Reflect, translate or rotate

Move the shapes as outlined.

Translate	Rotate $\frac{1}{4}$ turn clockwise	Reflect
1	2	3

Number and Algebra

SET 3 Multiples/number patterns

Complete each number pattern.

1	2	4				
2	3	6				
3	4	8				

Answer true or false.

4 Nine is a multiple of three. __________

5 Eleven is a multiple of two. __________

6 Twenty-four is a multiple of six. __________

7 Eighteen is a multiple of three, six and two. __________

8 Twenty-one is a multiple of three, seven and two. __________

9 The third multiple of four is fourteen. __________

10 The fourth multiple of five is twenty. __________

11 The fifth multiple of three is fifteen. __________

SET 4 Extension

1 ☐ ÷ 7 = 40

2 How much are 6 watches at $48 each?

3 435 cm – 40 cm

4 Tenths in $2\frac{1}{2}$

5 Average 150, 200, 100, 130

6 Double 237.

7 What number is halfway between 180 and 198?

8 What is half of $72.96?

9 $2.30 – 75c

10 Divide 240 by 3.

11 What is the difference between 144 and 40?

12 How much change did I receive from $20 if I spent $7.35?

13 If 4 cost $12, how much would 9 cost?

14 What is the sum of 3564 and 2035?

15 Centimetres in 3.7 m

16 Complete the grid.

18	9	36	27		72	45	
2			3	6		5	7

Measurement Cubic centimetres

Centicubes have been used to build these prisms. Record the volume of each prism. Colour the shape that has a different volume to the rest of the group.

1 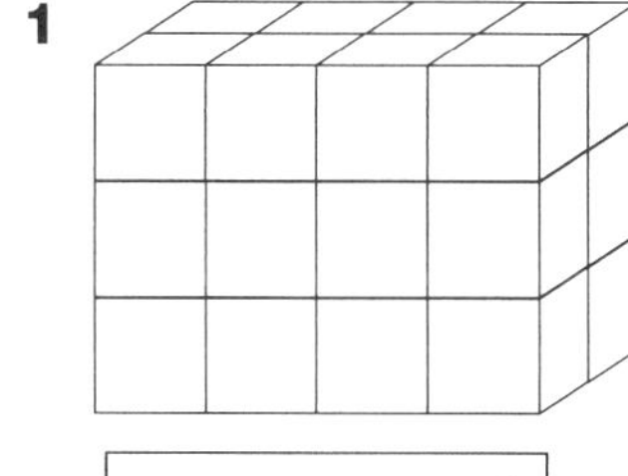

☐ cm^3

2 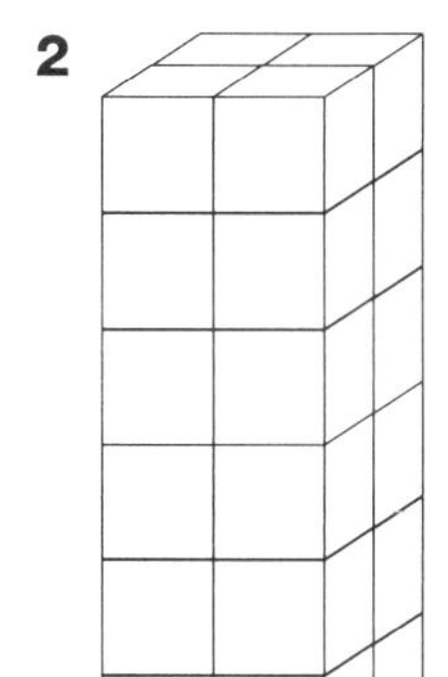

☐ cm^3

3 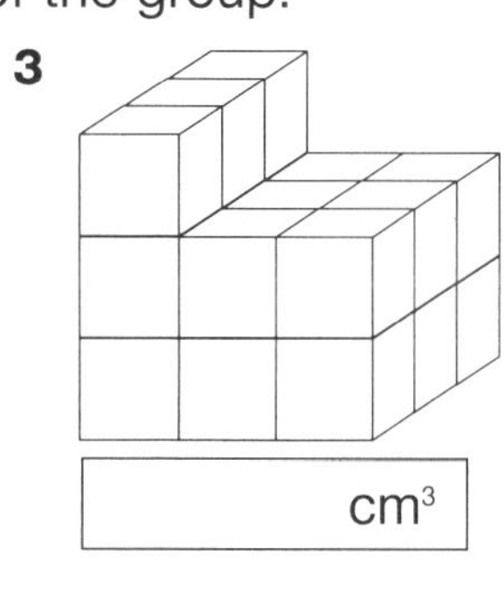

☐ cm^3

4 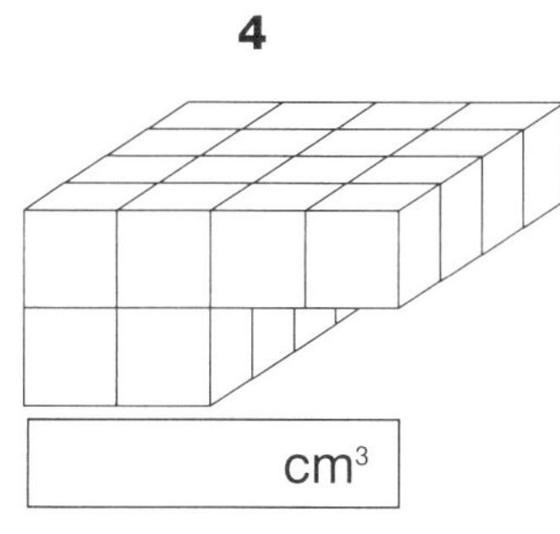

☐ cm^3

UNIT 4

Number and Algebra

SET 1 Basic

1 $7 + 4$

2 $8 + 9$

3 $19 - 5$

4 $11 - 8$

5 7×2

6 6×4

7 8×3

8 \$4.68 = ☐ c

9 $24 \div 4$

10 $30 \div 3$

11 What is the product of 7 and 4?

12 What is the sum of 8 and 6?

13 What is the value of 4 in 8416?

14 Divide 56 by 8.

15 

Ravi had 26 baseball cards and 52 football cards. How many cards did he have altogether?

☐ cards

SET 2 Rounding to 100 and 1000

Round each addend to the nearest 10 to estimate an answer for each addition.

1 $120 + 51$

2 $153 + 39$

3 $178 + 38$

4 $249 + 49$

Round each addend to the nearest 100 to estimate an answer for each addition.

5 $389 + 407$

6 $501 + 397$

7 $688 + 317$

8 $1099 + 299$

Round each addend to the nearest 1000 to estimate an answer for each addition.

9 $2013 + 5977$

10 $6878 + 3124$

11 $3750 + 3199$

12 $9999 + 4017$

13 $5699 + 5327$

Space Polygons

1 Describe these shapes.

a

____

____

____

____

b 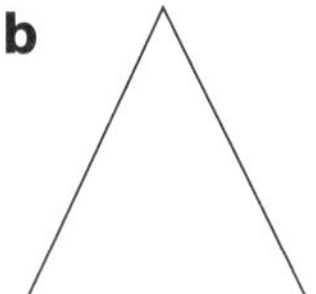

____

____

____

____

2 Colour the polygons.

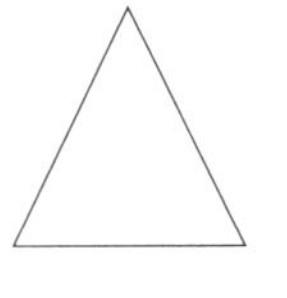  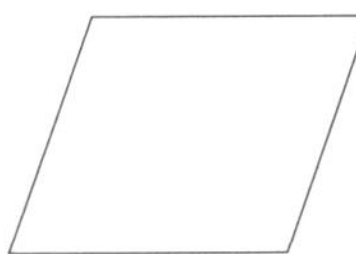

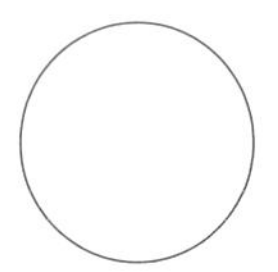 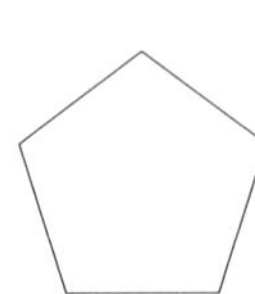 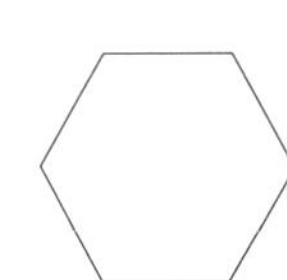

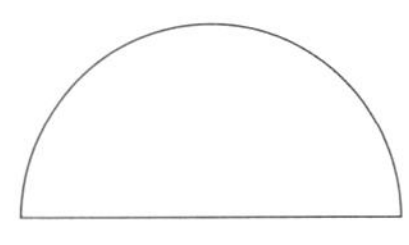  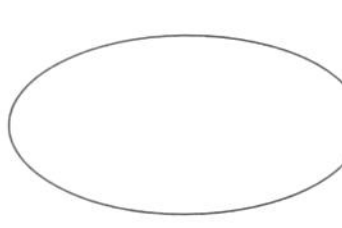

Number and Algebra

SET 3 Multiples of halves, quarters and eighths

One whole							
			$\frac{1}{2}$				
	$\frac{1}{4}$		$\frac{2}{4}$		$\frac{3}{4}$		
$\frac{1}{8}$	$\frac{2}{8}$	$\frac{3}{8}$	$\frac{4}{8}$	$\frac{5}{8}$	$\frac{6}{8}$	$\frac{7}{8}$	

Use the fraction wall to answer yes or no to the questions.

1 Is $\frac{1}{4}$ equivalent to $\frac{2}{8}$? ______

2 Is $\frac{1}{2}$ equivalent to $\frac{3}{4}$? ______

3 Is $\frac{3}{8}$ equivalent to $\frac{1}{4}$? ______

4 Is $\frac{3}{4}$ equivalent to $\frac{6}{8}$? ______

5 Is $\frac{1}{2}$ equivalent to $\frac{4}{8}$? ______

6 Is $\frac{4}{4}$ equivalent to one whole? ______

7 Complete the number line.

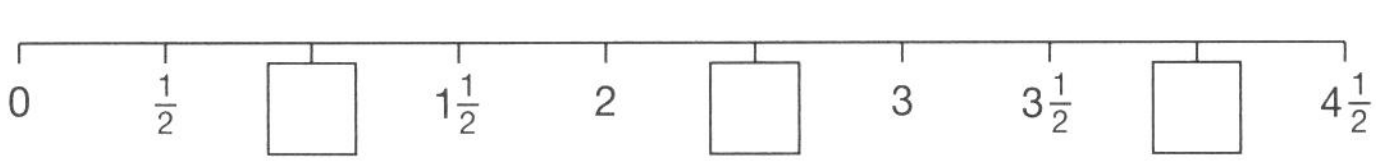

SET 4 Extension

1 (3 + 6) × 8

2 If 3 lollies cost 90c, how much would 20 cost?

3 \$867 – \$342

4 Which is larger, 0.7 or 25 hundredths?

5 How many grams in 2.7 kg?

6 How many minutes from 10 am to 2 pm?

7 Double 16 and add 83.

8 How many 5c lollies can be bought with \$1.95?

9 How much are 5 plants at \$9.50 each?

10 Divide 336 by 6.

11 What is the difference between 155 and 36?

12 How much is $3\frac{1}{2}$ kg of potatoes at \$3.50 per kg?

13 10% of 300

14 How many tens in 537?

15 One quarter of 260

Mathematical Reasoning

16 1, 1, 2, 3, 5, 8, 13, ____ ____ ____ ____

Leonardo Fibonacci discovered this number pattern. Write the next 4 Fibonacci numbers.

Measurement Estimating mass

Colour the masses which could balance a small boy on a see-saw.

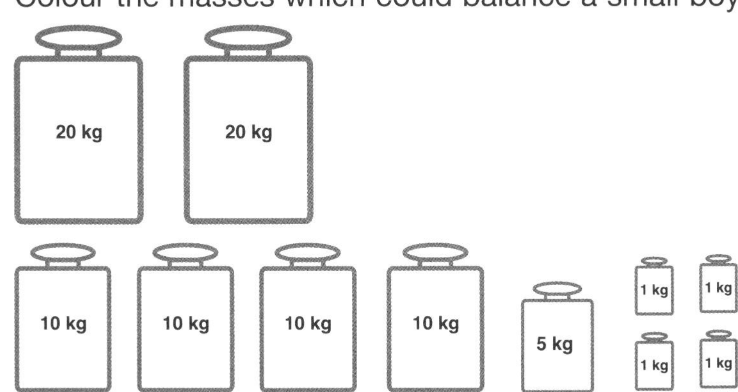

UNIT 5

Number and Algebra

SET 1 Basic

1 6 + 7

2 8 + 5

3 9 – 8

4 10 – 7

5 5 × 8

6 6 × 9

7 4 × 8

8 $6.71 = ☐ c

9 54 ÷ 9

10 63 ÷ 9

11 What is the product of 8 and 7?

12 What is the sum of 6 and 13?

13 Divide 40 by 8.

14 What is the value of 3 in 435?

15 

How many weeks will it take George to save $28 to pay for a T-shirt, if he saves $4 per week?
☐ weeks

SET 2 Multiplication strategies

Complete the facts.

		×4	×40
1	4		
2	6		
3	8		
4	10		
5	9		
6	7		

		×6	×60
7	4		
8	6		
9	8		
10	10		
11	9		
12	7		

13 How much are 5 dresses?

14 How much are 7 shorts?

15 How much are 9 T-shirts?

16 How much are 8 books and a dress?

Space Angles

Match each angle to its name.

Right	Acute	Reflex	Obtuse	Straight

Number and Algebra

SET 3 Problem solving

Distance from Brisbane

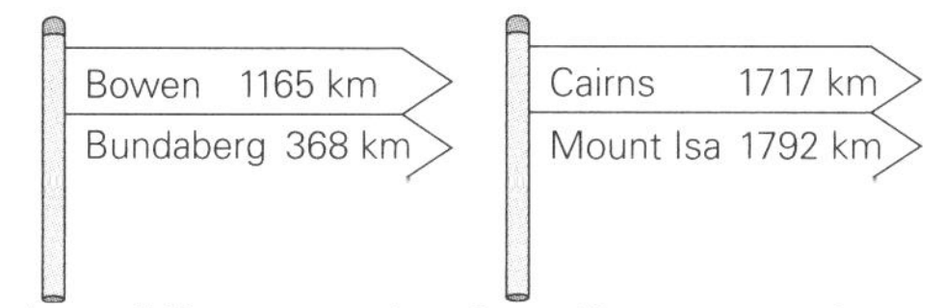

Find the difference in the distances between:

1 Bowen and Bundaberg
2 Mount Isa and Bundaberg
3 Cairns and Bowen
4 Mount Isa and Cairns
5 Cairns and Bundaberg
6 Mount Isa and Bowen

Mathematical Reasoning

7 Did Stephen lose $20 or not? ________

When he left home he had $125. When he came home he had $36 but thought he should have had $56.

These are the amounts he spent: $19, $20, $20, $30

SET 4 Extension

1 $26.54 = ☐ c
2 $5^2 + 1325$
3 Write the largest number you can using 3, 7, 6, 9, 1
4 5000 + 800 + 40 + 6
5 How many days in autumn?
6 If this Friday is the 13th, what is the date of the following Monday?
7 How much is 250 g at $8.60 per kilogram?
8 How many 25 cm lengths in $4\frac{1}{2}$ m?
9 37 more than 1986
10 If 7 cost $1.40, how much would 9 cost?
11 Are 27 and 35 multiples of 4?
12 How many 500 g bags are needed to fill a 12 kg bag?
13 How much is 5 kg at $7.80 a kilogram?
14 What is the difference between 240 and 137?
15 $(4 \times 10\,000) + (6 \times 1000) + (3 \times 100) + (9 \times 10) + 8$
16 What type of angle would be formed if the hands on a clock read 1 o'clock?

Statistics and Probability Likelihood

Impossible	Unlikely	Equal chance	Likely	Certain

How would you describe the chance of drawing a:

1 Purple marble from the bag? ________
2 Red marble from the bag? ________
3 Blue marble from the bag? ________
4 Yellow, red or green marble from the bag? ________
5 Yellow, red, green or purple marble from the bag? ________

Number and Algebra

SET 1 Basic

1 7 + 5

2 9 + 4

3 8 − 3

4 9 − 7

5 6 × 4

6 8 × 5

7 9 × 3

8 $6.80 = ☐ c

9 49 ÷ 7

10 48 ÷ 6

11 What is the product of 9 and 4?

12 What is the sum of 9 and 11?

13 Divide 42 by 6.

14 What is the value of 4 in 964?

15

A bag of lollies cost 25c. How much would 10 bags cost?

$ ☐

SET 2 Know 1, know 4

Write 4 facts you can derive from each array.

1

3 × 5 =
5 × 3 =
15 ÷ 3 =
15 ÷ 5 =

2

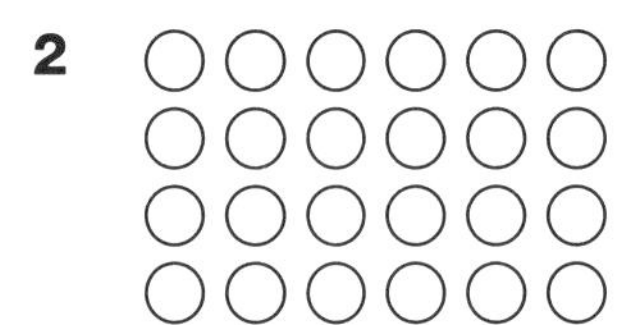

4 × 6 =
× =
÷ =
÷ =

3

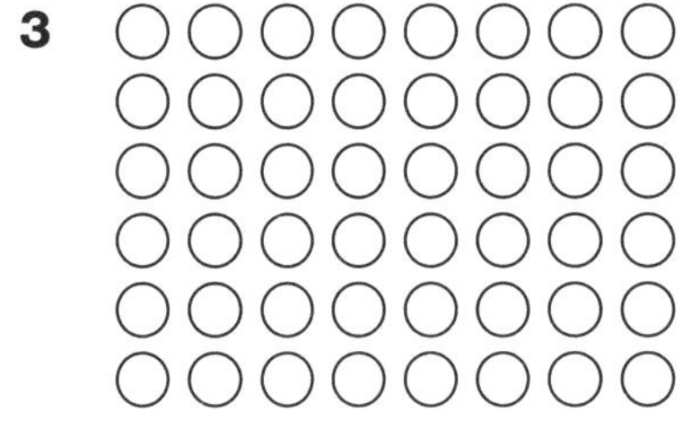

6 × 8 = 48
× =
÷ =
÷ =

Number and Algebra Division strategies

Use the number lines to show how repeated subtraction can solve division problems.

1 18 ÷ 3 = ______

0 1 2 3 4 5 6 7 8 9 10 11 12 13 14 15 16 17 18 19 20 21 22 23 24 25 26 27 28

2 25 ÷ 5 = ______

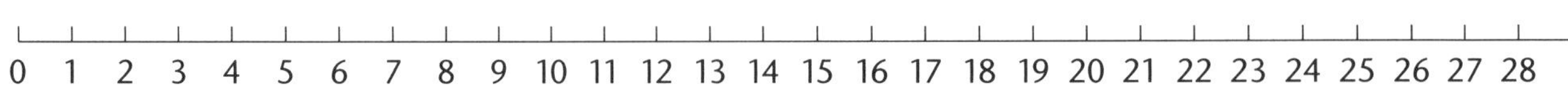

0 1 2 3 4 5 6 7 8 9 10 11 12 13 14 15 16 17 18 19 20 21 22 23 24 25 26 27 28

Number and Algebra

SET 3 Number patterns

Write rules to help you complete the patterns.

1 Rule:

23	27	31	35			

2 Rule:

2	7	12	17			

3 Rule:

59	56	53	50			

4 Rule:

1	2	4	8			

Mathematical Reasoning

5 Write the rule that will allow you to complete the grid describing the number of sticks used to create the pattern.

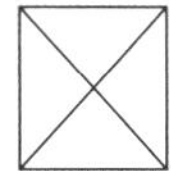
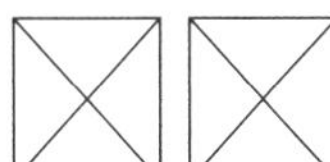
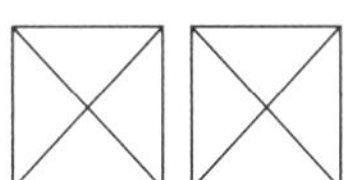
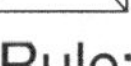

Rule:

Squares	1	2	3	4	5	6	7
Sticks	6	12					

SET 4 Extension

1 Estimate an answer to 345 plus 901.

2 $6^2 + 434$

3 $8\frac{1}{2}$ kg = ☐ g

4 $(72 \div 8) + 7$

5 Double 21 334.

6 If this Tuesday is the 21st, what is the date of the following Saturday?

7 \$6.70 – \$4.80

8 How many sides do 9 octagons have?

9 What is the product of (28 – 7) and 6?

10 How much are 5 combs at \$4.20 each?

11 0.73 = ☐ tenths + ☐ hundredths.

12 $\frac{3}{5}$ of \$80

13 How much is 9 kg of sausages at \$2.15 per kg?

14 What is the difference between 9999 and 999?

15 $(5 \times 10\,000) + (6 \times 1000) + (9 \times 100) + (6 \times 10) + 9$

16 If the large hand is on 11 and the small hand almost on the 9, what time is it?

Measurement 24-hour time/am and pm time

Express the times shown on the analog clocks in 24-hour time and using am and pm time.

1 

morning

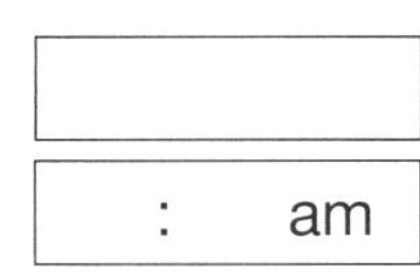

: am

2 

morning

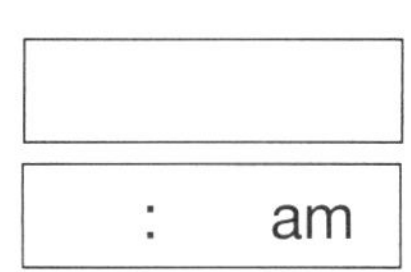

: am

3 

afternoon

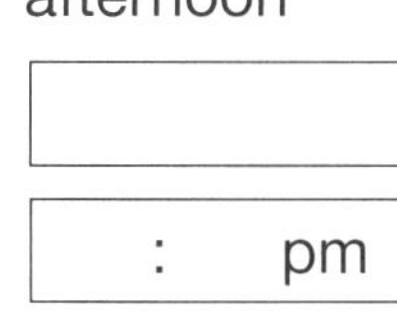

: pm

4 

evening

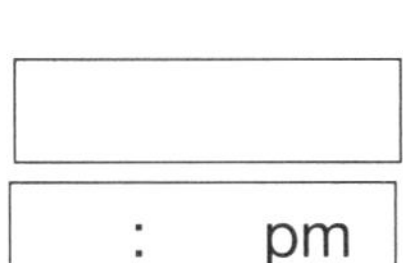

: pm

5 

evening

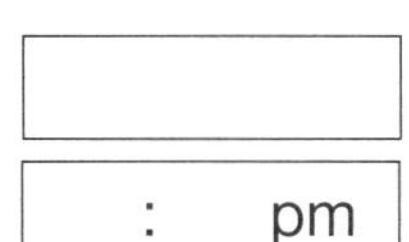

: pm

Number and Algebra

SET 1 Basic

1 6 × 7

2 8 × 5

3 25 – 13

4 43 – 5

5 18 + 9

6 16 + 5

7 25 ÷ 5

8 36 ÷ 6

9 20, 40, 60, ☐

10 What is the product of 7 and 5?

11 Divide 35 by 7.

12 How many minutes in 1 hour?

13 How much are 2 pears at 40c each?

14 Write the value of 9 in 93 562?

15

How many 50c packets of gum can Theo buy for $5? ☐

SET 2 3-digit multiplication

365 × 4
Think (300 × 4) + (60 × 4) + (5 × 4)
= 1200 + 240 + 20
= 1460

Split the number into hundreds, tens and ones in order to complete these multiplications.

1 324 × 3

2 452 × 4

3 591 × 2

4 259 × 2

Solve these multiplications.

5

	TH	H	T	O
		3	6	4
×				3

6

	TH	H	T	O
		4	9	6
×				5

7

	TH	H	T	O
		2	8	7
×				7

8

	TH	H	T	O
		6	0	9
×				5

Statistics and Probability Ordinal data

Rate these activities on a scale of 1–5 by shading a bubble for each option.

	Terrible (1)	Bad (2)	Average (3)	Good (4)	Excellent (5)
1 The last movie you watched.	○	○	○	○	○
2 The last meal you ate.	○	○	○	○	○
3 The last present you received.	○	○	○	○	○
4 The last holiday you went on.	○	○	○	○	○

Number and Algebra

SET 3 Equivalent fractions

Shade $\frac{1}{2}$ of each shape and write three equivalent fractions for $\frac{1}{2}$.

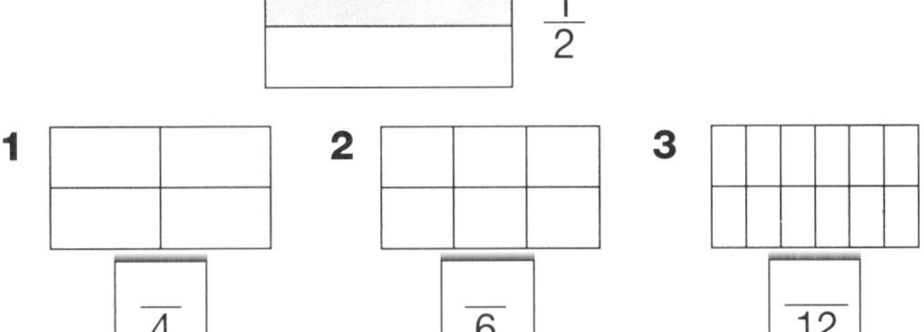

Shade $\frac{1}{4}$ of the top shape and create two other equivalent fractions.

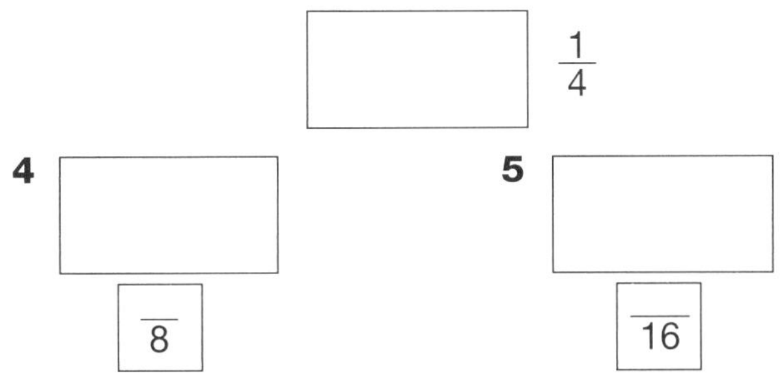

Shade the pair of equivalent fractions in each group.

6 $\frac{1}{3}$ $\frac{1}{2}$ $\frac{2}{4}$

7 $\frac{1}{3}$ $\frac{2}{4}$ $\frac{2}{6}$

8 $\frac{1}{2}$ $\frac{1}{4}$ $\frac{5}{10}$

9 $\frac{2}{4}$ $\frac{1}{5}$ $\frac{5}{10}$

10 $\frac{1}{5}$ $\frac{2}{10}$ $\frac{2}{3}$

11 $\frac{2}{8}$ $\frac{3}{8}$ $\frac{1}{4}$

SET 4 Extension

1. Write $\frac{81}{100}$ as a decimal.
2. How many $\frac{1}{4}$ s in $4\frac{3}{4}$?
3. $4^2 + 3$
4. Average 80, 60, 70 and 110.
5. How many 50 cm lengths make 10 m?
6. Write these numbers in ascending order 4127, 2356, 26 007.
7. How many 250 g packets make 10 kg?
8. Are 45 and 18 multiples of 9?
9. How much is 5 kg of potatoes at $3.50 per kilogram?
10. What is the sum of the odd numbers between 14 and 18?
11. What number is halfway between 1420 and 1430?
12. Write the set of factors of 40.
13. 420 minutes = ☐ hours
14. Write $2\frac{1}{2}$ as a decimal.

Mathematical Reasoning

15. In our class there are 5 girls to every 3 boys. How many girls are there if there are 32 children in the class altogether?

Measurement Metres and kilometres

Calculate the perimeter of these paddocks in metres and kilometres.

1

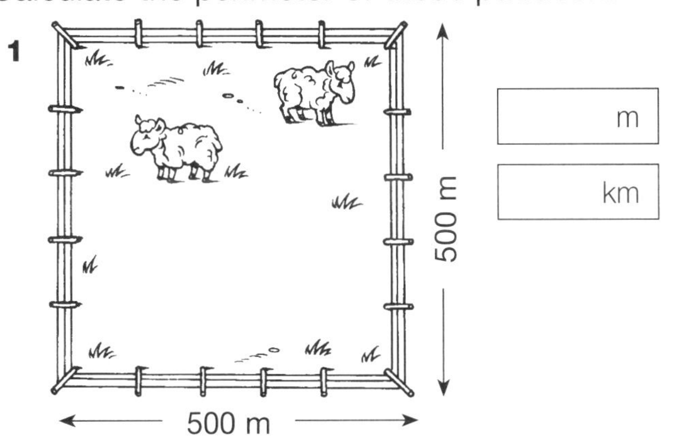

☐ m

☐ km

2

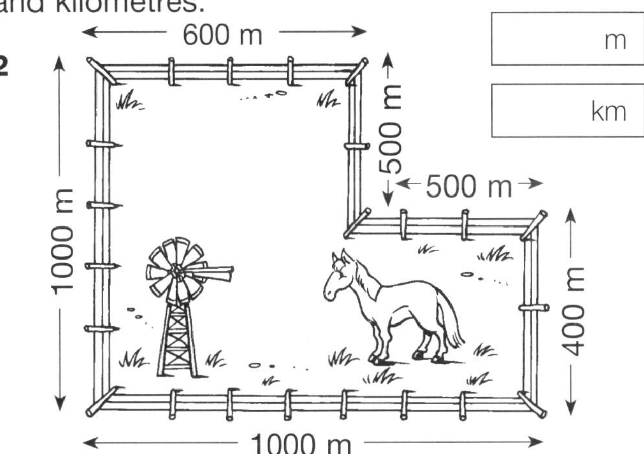

☐ m

☐ km

Number and Algebra

SET 1 Basic

1 25 ÷ 5

2 32 ÷ 4

3 16 + 5

4 27 + 6

5 32 – 5

6 4^2

7 What is the value of 6 in 2364?

8 7 × 4

9 6 × 5

10 $26.07 = ☐ c

11 What is the sum of 18 and 17?

12 What is the product of 9 and 7?

13 Divide 32 by 8.

14 What is the difference between 28 and 13?

15

SET 2 Subtraction strategies

Subtract the thousands first, then the hundreds followed by the tens and ones.

1 8959 – 6737

2 8348 – 7216

3 7433 – 5320

4 6528 – 4416

5 9486 – 7265

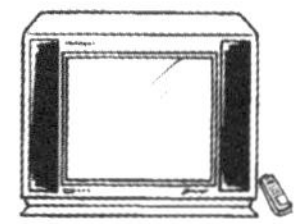

Calculate how much money Lee has left after buying each item, if she began the day with $4999.

	Purchase	Money left
6	Lounge	
7	TV	
8	Computer	

Space Compass points

What direction are these letters from 'E'?

1 A ________

2 B ________

3 C ________

4 D ________

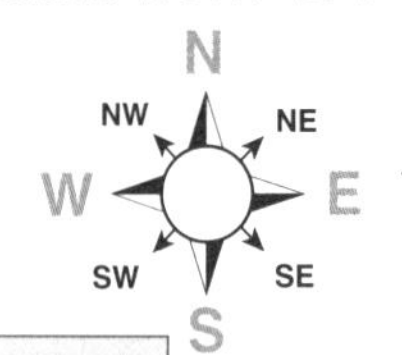

SCALE
1 cm = 100 m

Use the scale to calculate distances.

5 From A to B = ______ m

6 From B to C = ______ m

7 From C to D = ______ m

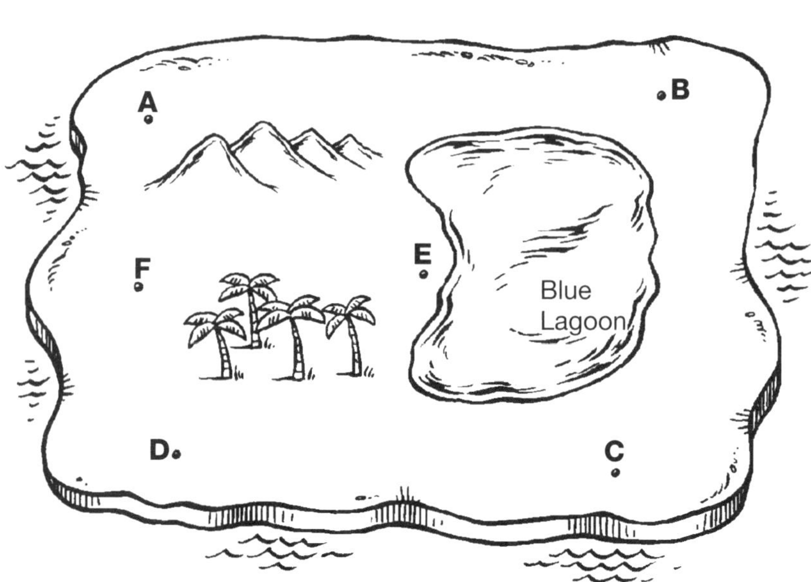

Number and Algebra

SET 3 Number patterns and rules

Apply the rules to complete the patterns.

1 Rule: × 3 + 1

21	22	23	24	25	26	27	28

2 Rule: – 8 × 5

79	80	81	82	83	84	85	86

3 Rule: × 4 – 2

2	4	6	8	10	12	14	16

4 Rule: ÷ 5 + 2

95	90	85	80	75	70	65	60

5 Work backwards to find the input numbers.

Rule: × 4 – 3

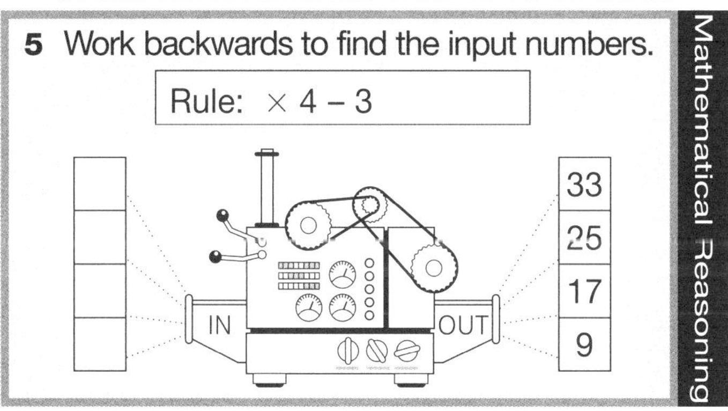

Mathematical Reasoning

SET 4 Extension

1 Add the even numbers between 13 and 17.

2 If 7 kg costs $35, how much would 9 kg cost?

3 $\frac{1}{10}$ = 0.1. True or false?

4 Which is larger, 1.0 or 0.1?

5 How many 0.75 m lengths in 3 m?

6 How many quarters in $5\frac{1}{2}$?

7 How many millimetres in 16.5 cm?

8 Write the smallest number you can using 4, 2, 3, 6, 7.

9 What shape are the sides of a triangular prism?

10 Write two thousand eight hundred and nine in figures.

11 Are 28 and 60 multiples of 9?

12 What time is 13 hours after 3 pm?

13 How much for 250 grams of nuts at $8 per kilogram?

14 Write 25 minutes past 7 am in digital form.

15 Write 30 027 in words.

Statistics and Probability Column graphs

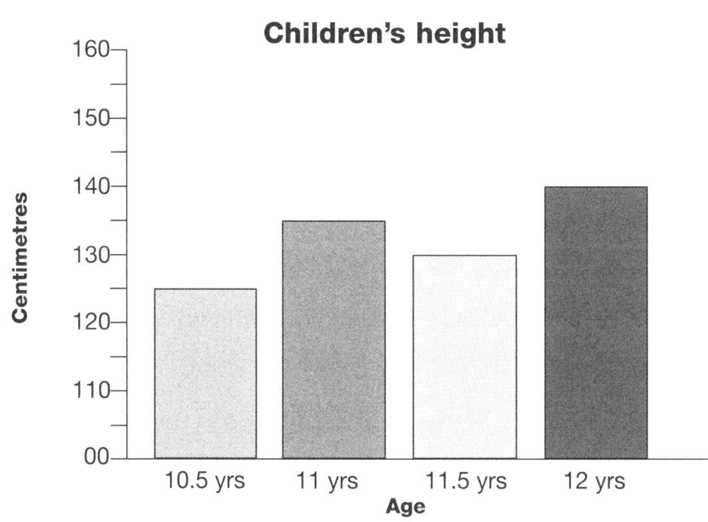

Jim measured the heights of 4 children in the playground and made this column graph.

1 How tall was the 10.5-year-old?

2 How tall was the 12-year-old?

3 How tall was the 11.5-year-old?

4 How much taller was the 12-year-old than the 10.5-year-old?

5 Is the second oldest person the second tallest?

UNIT 9

Number and Algebra

SET 1 Basic

1 6 × 4

2 21 ÷ 3

3 6 × 8

4 48 ÷ 6

5 4 × 10

6 ☐ × 5 = 25

7 6 + 6 + 6

8 (9 + 3) – 4

9 (7 + 7) – 3

10 (2 × 4) × 3

11 What is the product of 6 and 10?

12 What is the sum of 5, 4 and 8?

13 24 ÷ 6

14 18 ÷ 9

15 

There are 29 children in my class. If 17 are away sick how many are well?
☐ children

SET 2 Multiplication strategies

Halve one factor and double the other in order to find these products.

1 16 × 3

2 12 × 4

3 14 × 5

4 18 × 4

5 34 × 5

6 18 × 5

7 14 × 50

8 26 × 5

Use the double and double again strategy to multiply by 4.

	Question	Double	Double
9	1.6 × 4		
10	4.2 × 4		
11	2.5 × 4		
12	31.2 × 4		

Use the double, double and double again strategy to multiply by 8.

	Question	Double	Double	Double
13	5.2 × 8			
14	32.4 × 8			
15	14.3 × 8			
16	31.2 × 8			

Statistics and Probability Describing chance

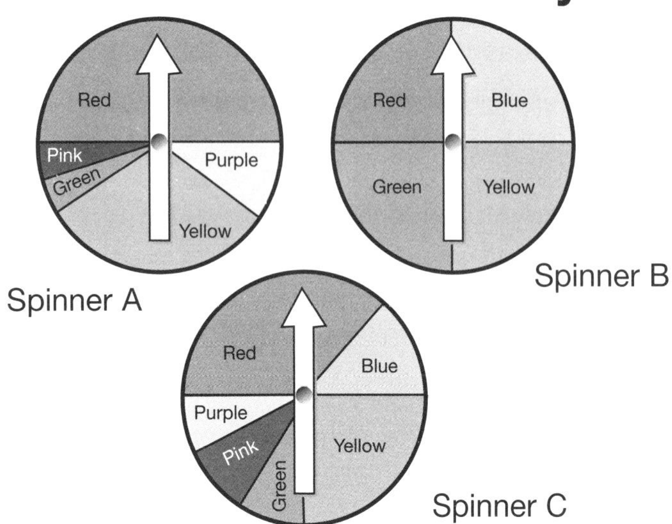

1 Which spinner has the most chance of landing on red?

2 Which spinner has the least chance of landing on green?

3 Which spinner has an equal chance of landing on red, blue, yellow or green?

4 Which spinner has an equal chance of landing on purple, pink or green?

5 Which spinners have the same chance of landing on yellow?

Number and Algebra

SET 3 Unit fractions of a group

1 Find $\frac{1}{4}$ of 20.

2 Find $\frac{1}{5}$ of 20.

3 Find $\frac{1}{2}$ of 80.

4 Find $\frac{1}{3}$ of 27.

5 Find $\frac{1}{8}$ of 40.

6 Find $\frac{1}{10}$ of 100.

7 Find $\frac{1}{4}$ of 240 sheep.

8 Find $\frac{1}{5}$ of 500 goats.

9 Find $\frac{1}{8}$ of 1600 cakes.

10 Find $\frac{1}{5}$ of 5000 rabbits.

11 Below each person write their age.

_____ _____ _____ _____ _____

Tom is $\frac{1}{4}$ of Will's age.

Laura is $\frac{1}{2}$ of Will's age.

Kate is $\frac{1}{6}$ of Susan's age.

Kate, who is 6, is $\frac{1}{8}$ of Will's age.

SET 4 Extension

1 Estimate an answer to 287 + 409.

2 12.5 + 1.3

3 ☐ kg = 0.75 tonne

4 How much is 5 L at $4.50 per litre?

5 940 m + 59 m

6 (48 ÷ 6) + 463

7 $\frac{3}{8}$ of $960

8 A cricketer scored 4, 19, 27, 6. What was his average?

9 1302, 1332, 1362, ☐, ☐

10 How much is 500 g at $37 per kilogram?

11 How much change did I receive from $20 if I spent $1.65?

12 $107 + $52 + $218

13 $79.05 = ☐ c

14 Four girls have $60 between them. How much has each girl got if:

Anna has twice as much as Bree.	$
Bree has 3 times as much as Deni.	$
Cath has $\frac{1}{6}$ of the total amount.	$
Deni has $\frac{1}{12}$ of the total amount.	$

Mathematical Reasoning

Space Shapes (translate, rotate and reflect)

Follow the instructions to make a pattern.

	Reflect	Rotate 90°	Reflect	Rotate 90°	Reflect

UNIT 10

Number and Algebra

SET 1 Basic

1 18 + 3

2 21 − 3

3 6 × 9

4 18 ÷ 3

5 20 ☐ 5 = 4

6 15 ☐ 3 = 5

7 20 ☐ 6 = 26

8 What is the product of 9 and 12?

9 22 ☐ 8 = 14

10 What is the sum of 18 and 10?

11 How many minutes in 2 hours?

12 How much are 3 lollies at 5c each?

13 $\frac{1}{2}$ of 100

14 Divide 12 by 12.

15 

Each milk crate holds 16 L of milk. How many litres would 6 milk crates hold?

☐ L

SET 2 5-digit addition

1
```
  24 655
+ 17 409
```

2
```
  38 080
+ 24 533
```

3
```
  43 720
+ 15 592
```

4
```
  48 364
+ 39 783
```

5
```
  55 555
+ 37 680
```

6
```
  16 356
+ 63 562
```

7
```
  16 341
  23 837
+ 17 601
```

8
```
  21 007
  35 632
+ 12 555
```

9 Sophia bought a car for $43 990, roof racks for $305, window tinting for $430 and tow bar for $385. How much did she spend? 

Space Drawing prisms

Repeat each prism on the isometric dot paper. Each has been started for you.

1

2 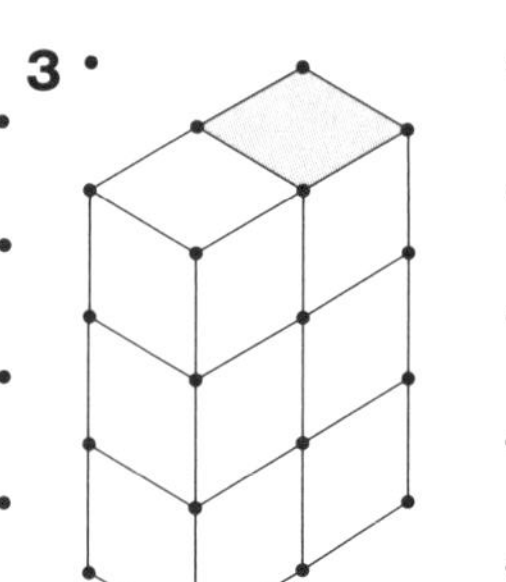

3 

Number and Algebra

SET 3 Multiples

Answer true or false.

1 12 is a multiple of 3.

2 10 is a multiple of 2.

3 15 is a multiple of 3.

4 24 is a multiple of 3.

5 12 is a multiple of 4.

6 12 is a multiple of 5.

7 32 is a multiple of 4.

8 24 is a multiple of 6.

9 24 is a multiple of 7.

10 36 is a multiple of 8.

Write the next five multiples of:

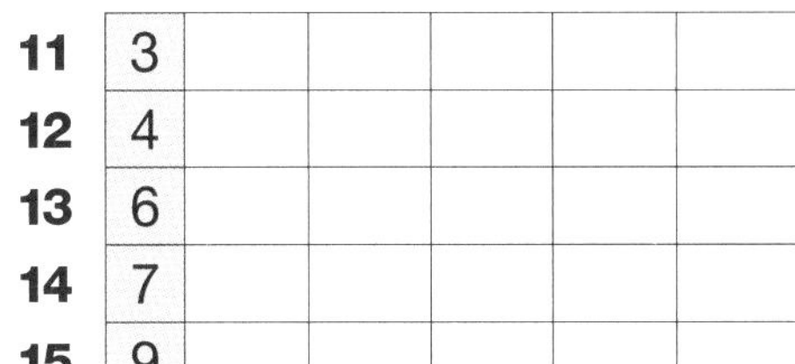

11	3					
12	4					
13	6					
14	7					
15	9					

SET 4 Extension

1 0.2 = 2 hundredths. True or false?

2 $5^2 + 4$

3 How many $\frac{1}{4}$ s in $2\frac{1}{4}$?

4 Write the factors for 45.

5 How many 25 cm lengths can be cut from 2 m?

6 How many surfaces has a pentagonal prism?

7 27×100

8 Are 18 and 50 multiples of 8?

9 How much is $2\frac{1}{2}$ kg of bacon at $6.40 per kilogram?

10 Average 10, 14 and 18

11 What is the sum of multiples of 3 between 11 and 16?

12 Write these numbers in descending order 27 301, 28 963, 35 427.

13 Which is larger, $\frac{3}{4}$ or 0.5?

14 How much are 9 apples at 25c each?

15 Write 20 206 in words.

____

____

Measurement Square centimetres

Calculate the area of these shapes.

1

Area = ☐ cm^2

2

Area = ☐ cm^2

3

Area = ☐ cm^2

UNIT
11

Number and Algebra

SET 1 Basic

1 17 + 5

2 26 ☐ 4 = 30

3 30 – 15

4 24 ☐ 8 = 16

5 7×5

6 26 ☐ 2 = 13

7 $24 \div 3$

8 9 ☐ 3 = 27

9 195, 200, 205, ☐

10 1 minute = ☐ seconds

11 What is the product of 9 and 7?

12 What is the difference between 23 and 7?

13 How much are 7 lollies at 8c each?

14 How many years in 1 century?

15 

Zorba had 7 boxes of chocolates with 30 chocolates in each. How many chocolates did Zorba have?

☐ chocolates

SET 2 Division strategies

Write a multiplication fact to check each division.

1 $21 \div 3 = 7$ ✓ $7 \times 3 = 21$

2 $49 \div 7 = 7$ ☐ ______

3 $54 \div 6 = 9$ ☐ ______

4 $54 \div 8 = 7$ ☐ ______

5 $36 \div 4 = 8$ ☐ ______

6 $81 \div 9 = 9$ ☐ ______

7 $56 \div 8 = 6$ ☐ ______

8 $45 \div 5 = 9$ ☐ ______

9 $42 \div 6 = 7$ ☐ ______

10 $3\overline{)72}$

11 $6\overline{)96}$

12 $7\overline{)79}$

13 $2\overline{)54}$

14 $5\overline{)90}$

15 $5\overline{)83}$

16 $3\overline{)78}$

17 $7\overline{)77}$

18 $4\overline{)66}$

Statistics and Probability Column graphs

Fletcher has made a graph to display the number of points he scored in the basketball tournament.

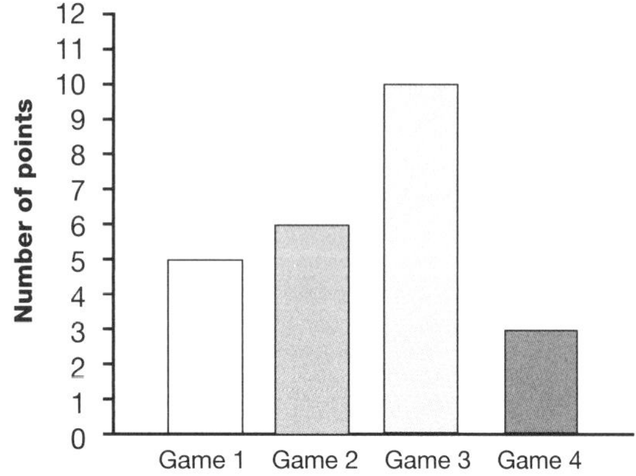

1 How many more points did he score in Game 2 than in Game 1? ☐

2 What was the difference between his highest score and his lowest score? ☐

3 How many points did he score altogether? 

4 In one game he only played half a game. Which game do you think it was? ☐

Number and Algebra

SET 3 2-digit multiplication and estimation

Estimate an answer to these multiplications by rounding off to the nearest 10.

1 29×3

2 49×5

3 99×2

4 31×5

5 42×3

6 39×5

7 21×7

8 32×9

9 48×6

10 29×7

11 38×8

12 42×6

13 51×9

14 37×5

15 41×8

SET 4 Extension

1 $4\frac{1}{4}$ m = ☐ centimetres

2 Are 27 and 45 multiples of 9?

3 How many 750 g packets make $4\frac{1}{2}$ kg?

4 $\frac{7}{8}$ of 168

5 $6^2 + 5$

6 How much are 5 boxes of chocolate at $2.20 each?

7 37×8

8 How much is 4 kg of steak at $2 per $\frac{1}{2}$ kilogram?

9 How many 60 cm lengths in 3 m?

10 Write these numbers in ascending order 53 224, 20 206, 7496.

11 What is the perimeter of a triangle with three 14 cm sides?

12 How many hundreds in 4230?

Mathematical Reasoning

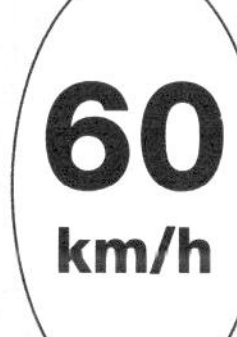

13 What is the maximum distance a driver could go in $2\frac{1}{2}$ hours if they kept to the speed limit?

Space Prisms

Colour each prism and its net the same colour.

1 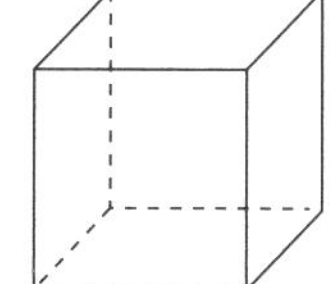

2 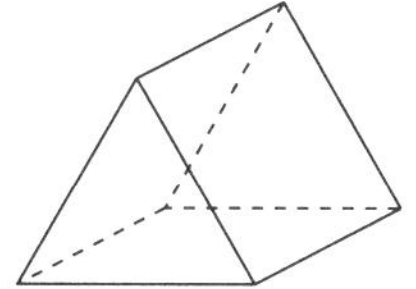

3 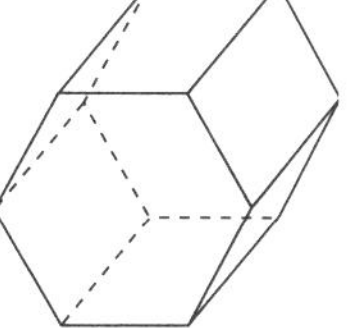

4 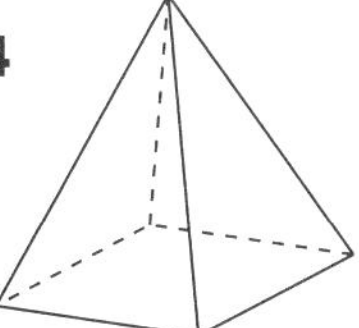

5 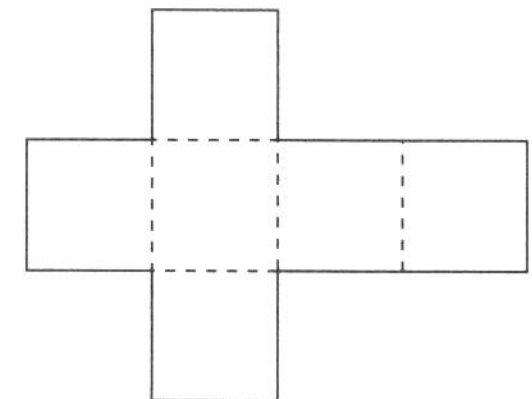

6 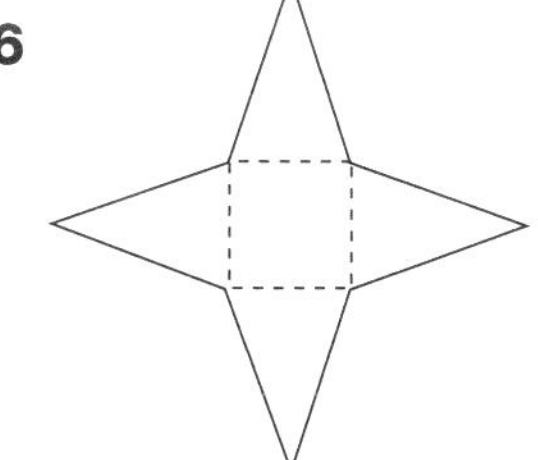

7 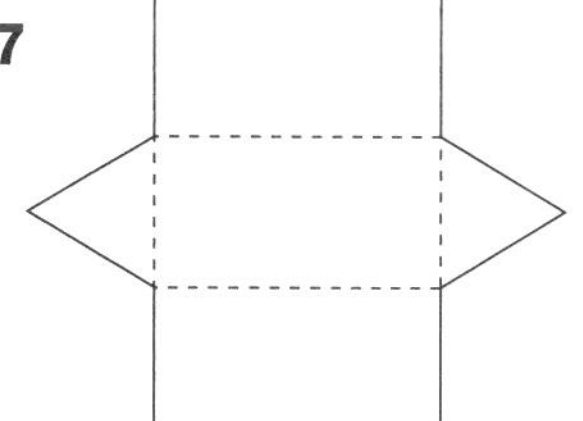

8 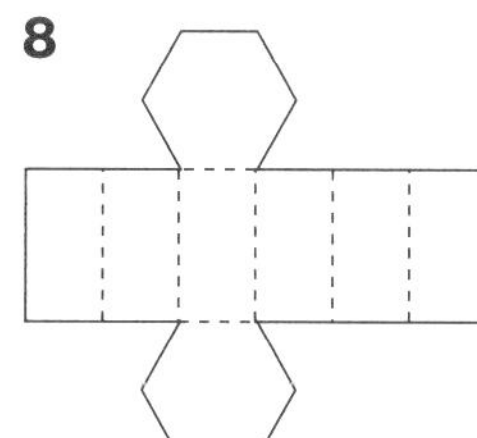

Number and Algebra

SET 1 Basic

1 4×11

2 $\$3 \times 4$

3 68 = ☐ tens + ☐ ones

4 \$4 + \$8 + \$5

5 5^2

6 200 + 30 + 9

7 \$9.70 – \$2

8 7^2

9 How much are 2 brushes at \$1.30 each?

10 4×1000

11 Divide 27 by 3.

12 Multiply 6 by 5.

13 $\$4 \times 10$

14 Add 40, 10 and 50.

15

SET 2 5-digit subtraction

Find the differences in prices between these cars.

1

van	
car	

2

4WD	
ute	

3

station wagon	
ute	

4

ute	
car	

5

4WD	
station wagon	

6

station wagon	
car	

Measurement Cubic centimetres

How many cubic centimetres (centicubes) would be needed to fill each box?

1

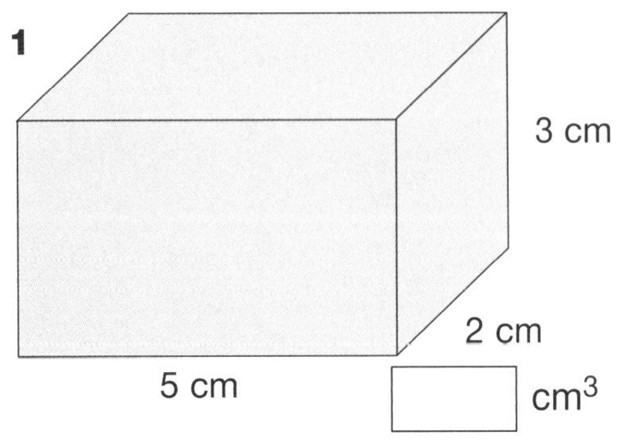

2

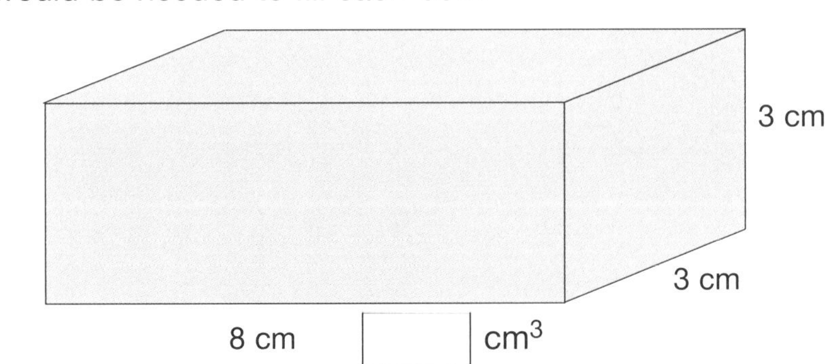

Number and Algebra

SET 3 Factors

Find four factors for each number.

	Number	Factors
1	20	
2	16	
3	12	
4	40	
5	32	
6	30	

7 Is 3 a factor of 12?

8 Is 8 a factor of 32?

9 Is 6 a factor of 27?

10 Is 6 a factor of 18?

11 Find all the factors for 24.

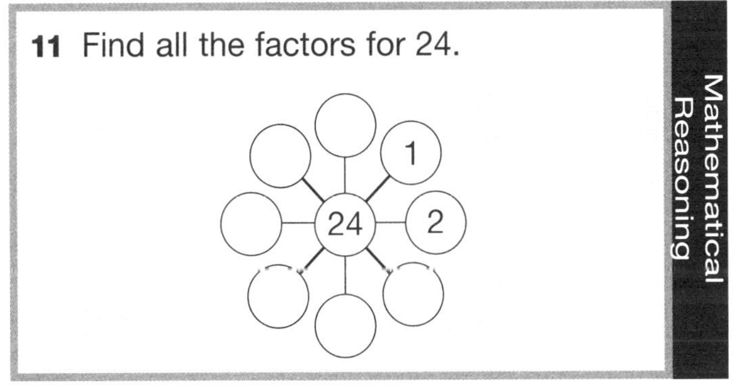

SET 4 Extension

1 $\square \div 8 = 7$

2 \$2.30 + \$1.25 + \$1.48

3 $(\frac{3}{5} \times 20) + 250$

4 What is 57 more than 1374?

5 What is the perimeter of a square with 17 cm sides?

6 How much change did I receive from \$20 if I spent \$4.78?

7 Round 6538 to the nearest thousand.

8 If 7 cost \$2.80, how much for 9?

9 Share \$4.80 among 3 people.

10 Write the multiples of 7 between 13 and 50.

11 How much are 6 pens at \$7.10 each?

12 How many tenths in $7\frac{1}{2}$?

13 How many minutes from 9:55 to 12:30?

14 $(8 \times 10^2) + (6 \times 10)$

15 How many hundreds in 15 287?

16 How many 15 mL medicine bottles can be filled by a 300 mL jug?

17 Round to the nearest whole kilometre to answer 15.15 km + 27.89 km.

Space Translate, rotate and reflect

Translate, rotate and reflect the given shape.

Shape	Translate	Rotate	Reflect
N			

UNIT 13

Number and Algebra

SET 1 Basic

1 3, 6, 9, ☐, 15

2 10×9

3 8×5

4 $16 \div 4$

5 $28 \div 4$

6 $(2 + 5) \times 3$

7 $(3 + 3) \times 4$

8 600 + 8

9 100 – 10

10 $5^2 - 2$

11 How much are 7 pens at $1.10 each?

12 What is one quarter of 40?

13 $6.21 = ☐ c

14 $10 × 3

15 Mark planted 7 rows of flowers with 8 in each row. How many flowers did he plant?

☐ flowers

SET 2 3-digit multiplication

Complete these algorithms.

1 163×5

2 281×3

3 494×6

4 295×6

5 437×7

6 826×3

7 A bakery cooks 240 biscuits each day. How many biscuits are baked in 7 days?

8 Willow ran 25 km per week for 8 weeks. What was the total distance she ran?

9 Thomas bought 8 chairs at $175 each. How much did he spend?

Space Nets of 3D objects

Circle the prism that belongs to the net.

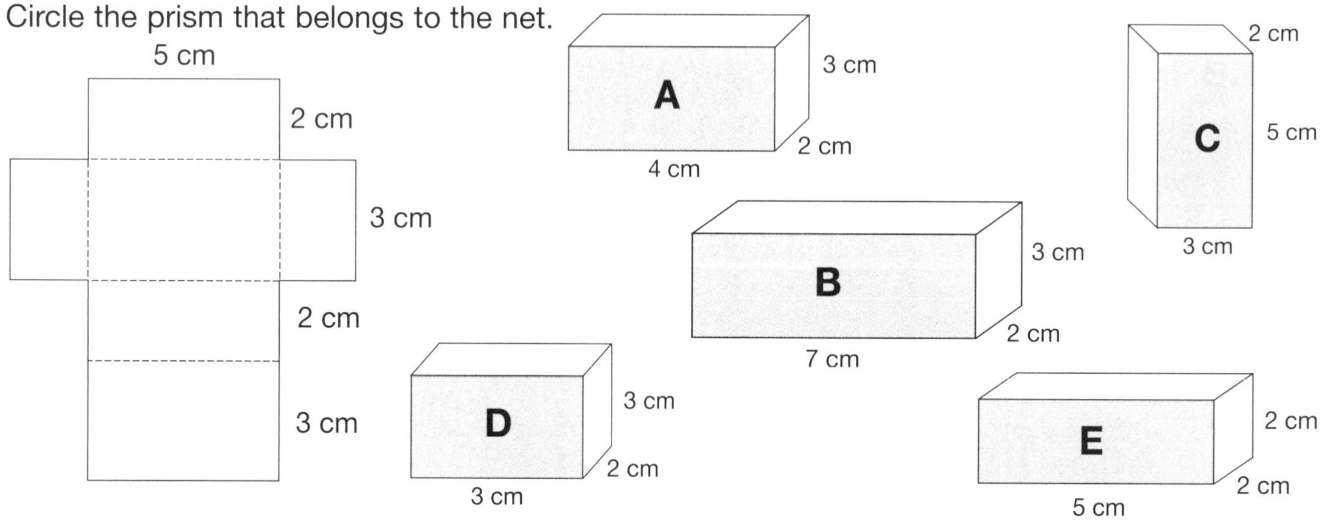

Number and Algebra

SET 3 Order of operations

1 $3 + 6 \times 2 =$

2 $(3 + 6) \times 2 =$

3 $2 \times 7 + 5 =$

4 $2 \times (7 + 5) =$

5 $3 \times 5 \times 2 =$

6 $2 \times 3 \times 4 + 7 =$

7 $3 + 2 \times 7 + 3 =$

8 $30 - 3 \times 5 =$

9 $(10 - 7) \times 5 - 6 =$

10 $40 - 7 \times (8 - 6) =$

Write true or false.

11 $3 \times 7 + 9 = 30$ ________

12 $3 \times (7 + 9) = 30$ ________

13 $5 \times 8 + 16 = 56$ ________

14 $5 + 3 \times 9 + 7 = 39$ ________

15 $10 + 5 \times 6 + 13 = 77$ ________

16 $80 - 3 \times 10 + 7 = 57$ ________

17 $3 \times 9 - 10 - 8 = 10$ ________

18 $250 - 9 \times 10 - 70 = 190$ ________

19 $300 - 7 \times 9 + 63 = 237$ ________

20 $400 - 7 \times (5 + 6) = 223$ ________

Mathematical Reasoning

21 (□ + △) × 5 = ⬡60

Supply your own numerals to complete this number sentence. You are not allowed to use a numeral more than once.

SET 4 Extension

1 25×1000

2 What is the sum of 1323 and 506?

3 List the factors of 42.

4 $(\frac{7}{8} \times 64) + (\frac{3}{4} \times 120)$

5 What is the difference between 357 and 203?

6 How much is $3\frac{1}{4}$ kg of meat at $12 per kilogram?

7 What is the value of 7 in 2.74?

8 Round off 38 126 to the nearest thousand.

9 Which is larger, 7^2 or 10×9?

10 $4000 + 600 + 50 + 8$

11 How much are 6 apples at 3 for $1.25?

12 6:25 pm + 45 minutes

13 If a square has a perimeter of 24 cm, what is the length of each side?

Mathematical Reasoning

14 Tina bought 10 packets of chocolate biscuits. Can you calculate the cost faster than using a calculator?

Biscuits	$3.95
Biscuits	$3.95
Biscuits	$3.95
Biscuits	$3.95
Biscuits	$3.95
Biscuits	$3.95
Biscuits	$3.95
Biscuits	$3.95
Biscuits	$3.95
Biscuits	$3.95

Measurement 24-hour time

Write 24-hour times for each program.

1 'The Last Frontier' □

2 'The Midday Show' □

3 'Days of Our Lives' □

4 'Who's the Boss?' □

5 'A Current Affair' □

6 What program starts at 19:30?

7 What program starts at 22:30?

6:00	**Tennis** – The US Open.
9:30	**The Last Frontier** – Wildlife.
10:00	**Romper Room** – For kids.
10:30	**Police Story** – (PGR, Final).
11:30	**Wheel of Fortune** – Game.
12:00	**The Midday Show** – Topical.
1:30	**Days of Our Lives** (PGR).
2:35	**The Young and the Restless.**
3:30	**Focus on Living** – Religious.
4:00	**Who's the Boss?** – (Return).
4:30	**Bush Beat** – For children.
5:00	**Happy Days** – Comedy series.
5:30	**Sale of the Century** – (*S).
6:00	**News, Sport and Weather.**
7:00	**A Current Affair** – Topical.
7:30	**A Country Practice** – (*S, PGR).
8:30	**Twin Peaks** – (PGR). The identity of Laura Palmer's killer is revealed; Leland tries to discourage Maddy from going home.
9:30	**MacGyver** – (PGR). Drama.
10:30	**Night Court** – (PGR). Comedy.
11:00	**Motorcycle Racing** – Le Mans 500cc Grand Prix.
12:00	**Close.**

UNIT 14

Number and Algebra

SET 1 Basic

1 5×8

2 42 ☐ 7 = 6

3 9 ☐ 9 = 81

4 465, 455, ☐, 435

5 6 chocolate bars at 60c each

6 One half of 90

7 $36 \div 9$

8 Multiply 7 by 11.

9 What is the sum of 60 and 25?

10 How many eggs in 2 dozen?

11 $10^2 + 7$

12 What is the difference between 53 and 39?

13 What is the value of 2 in 60 208?

14 $(6 + 3) \times 5$

15 How many 80c ice-creams can you buy for $4.00? ☐

SET 2 Division strategies

1

÷	360	120	480	240	600	300	540
6							

Use the halve and halve again strategy to divide by 4.

	Question	Halve	Halve
2	40 ÷ 4	20	10
3	24 ÷ 4		
4	64 ÷ 4		
5	200 ÷ 4		
6	88 ÷ 4		
7	160 ÷ 4		

Use the halve, halve again and halve again strategy to divide by 8.

	Question	Halve	Halve	Halve
8	56 ÷ 8	28	14	7
9	80 ÷ 8			
10	200 ÷ 8			
11	160 ÷ 8			
12	96 ÷ 8			

Statistics and Probability Chance experiment

1 Which colour marble has the most chance of being pulled out of the bag?

2 Which marble has the least chance?

3 Is it more likely that yellow will be picked than green?

4 Is it more likely that pink will be picked than yellow?

5 Which colour has an even chance at being picked from the bag?

R = red G = green Y = yellow P = pink

Number and Algebra

SET 3 Mixed numerals

Draw a line to match the improper fractions to the mixed numbers.

1 $\frac{6}{5}$	$1\frac{4}{5}$
2 $\frac{12}{10}$	$1\frac{1}{5}$
3 $\frac{14}{8}$	$1\frac{2}{10}$
4 $\frac{9}{5}$	$1\frac{6}{8}$
5 $\frac{17}{10}$	$1\frac{7}{10}$
6 $\frac{11}{8}$	$1\frac{3}{6}$
7 $\frac{9}{6}$	$1\frac{3}{8}$

These fractions have been added to give improper fractions. Convert each answer into a mixed number.

8 $\frac{6}{8} + \frac{5}{8} = \frac{11}{8} =$

9 $\frac{4}{5} + \frac{3}{5} = \frac{7}{5} =$

10 $\frac{7}{10} + \frac{9}{10} = \frac{16}{10} =$

11 $\frac{4}{6} + \frac{4}{6} = \frac{8}{6} =$

12 $\frac{3}{8} + \frac{7}{8} = \frac{10}{8} =$

13 $\frac{8}{12} + \frac{10}{12} = \frac{18}{12} =$

SET 4 Extension

1. Round 18 627 to the nearest 1000.
2. ☐ ÷ 9 = 11
3. $6.50 × 100
4. List the factors of 32.
5. How many hours from 6 am to 3:30 pm?
6. How much change do I receive from $25 if I spent $22.45?
7. How many months in $3\frac{1}{3}$ years?
8. Which is larger, $\frac{1}{2}$, 80% or $\frac{60}{100}$?
9. If 5 cost $2.00, how much do 2 cost?
10. If a square has an area of 36 cm^2, what is the length of the base?
11. How many 400 g bags in 6 kg?
12. 6×5^2
13. Share $6.30 among 7 people.
14. How many 20 cm lengths in 1.6 m?
15. It took 5 minutes to run one lap. How many laps could be run in 2 hours?

Mathematical Reasoning

16. Give the dimensions in metres of a square whose perimeter and area are the same number.

Measurement Perimeter

Measure the following perimeters in centimetres.

1

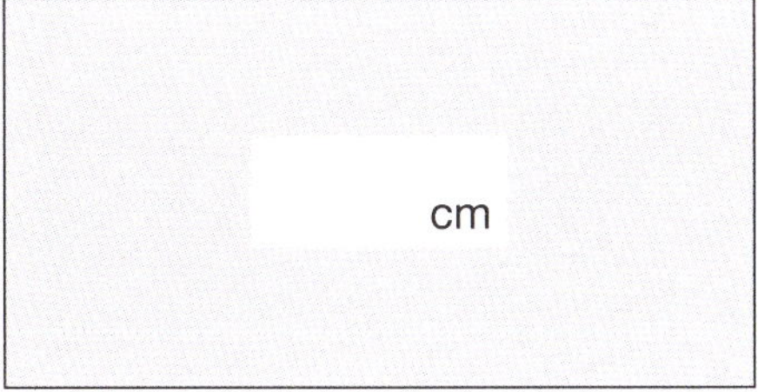

cm

2

cm

3

cm

Number and Algebra

SET 1 Basic

1 1290, 1295, 1300, ☐

2 27 – 6

3 35 ☐ 7 = 42

4 18 + 9

5 50 ☐ 30 = 20

6 6 × 9

7 48 ÷ 8

8 6 ☐ 7 = 42

9 18 ☐ 6 = 3

10 What is the product of 7 and 8?

11 How many millilitres in 1 L?

12 5 pencils at 5c each

13 What is the value of 7 in 37 256?

14 How many years in a decade?

15 

Wayan picks 23 coconuts an hour. How many will he pick in 7 hours? ☐ coconuts

SET 2 3-digit multiplication

1 236×3

2 182×4

3 294×5

4 592×6

5 432×3

6 642×7

7 365×4

8 742×8

9 482×3

10 There are 8 cans of soup in a box. If each can weighs 415 g, what is the mass of the box? 

Mathematical Reasoning

11 Chris earned \$24. If Emma earned $\frac{1}{3}$ of that amount, which was equal to $\frac{1}{6}$ of Mark's earnings, how much did Emma and Mark each earn?

Emma = \$ ______ Mark = \$ ______

Statistics and Probability Line graphs

Create a line graph from the data in the table.

Time	6 am	8 am	10 am	noon	2 pm	4 pm
Temperature	14°C	16°C	20°C	24°C	24°C	22°C

Estimate the temperature at 11 am? ☐

Temperature for 6 am to 4 pm

Temperature

Time

Number and Algebra

SET 3 Using factors

Use the square and rectangular shapes to describe the pairs of factors for each number.

1

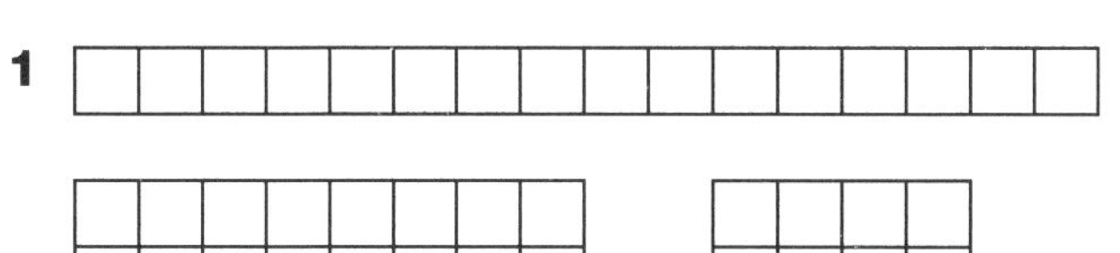

Factors of 16:

____

2

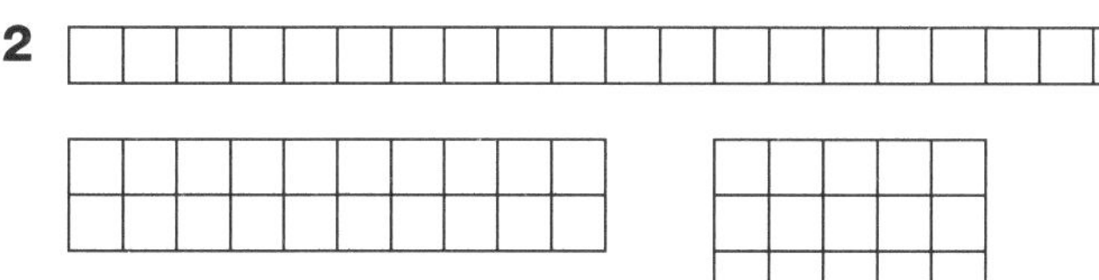

Factors of 20:

____

SET 4 Extension

1 $\frac{3}{10}$ of 200

2 Round 21 350 to the nearest 1000.

3 9999 + 801

4 $8\frac{1}{2}$ km = ☐ m

5 Write the numeral for twenty-seven thousand, two hundred and twenty-six.

6 96 × 70

7 List prime numbers between 20 and 30.

8 Which is larger, 2.5 or $2\frac{1}{10}$?

9 How many minutes from 3:25 to 7:00?

10 Which is larger, $2\frac{1}{3}$ or $2\frac{1}{4}$?

11 What is the value of 9 in 3.79?

12 What is the perimeter of a rectangle with length 7 cm and width 3 cm?

13 How many 35 cm lengths in 7 m?

14 Average of 15, 30, 45, 70

15 How much is 500 g of meat at $7 per kilogram?

16 Alexis walks 2 km in 10 minutes. How far will she walk in 65 minutes?

Measurement Square metres

Calculate the areas of these shapes using the scale.

SCALE
1 cm = 1 m

1

Blackboard ___ m^2

2

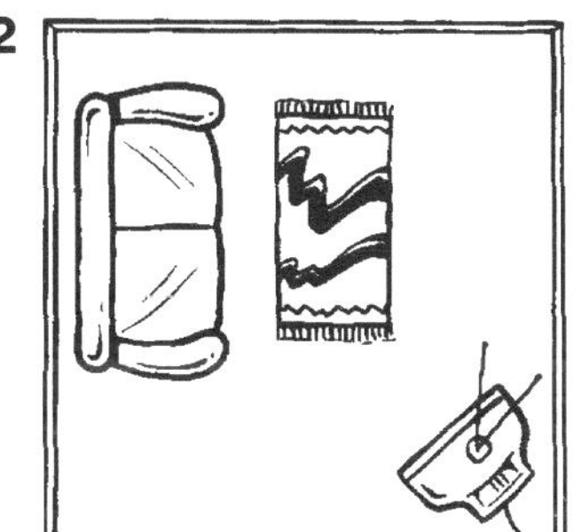

Lounge room ___ m^2

3

Swimming pool ___ m^2

Number and Algebra

SET 1 Basic

1 4, 8, ☐, ☐, 20

2 9 × 7

3 8 × 6

4 20 ÷ 5

5 30 ÷ 6

6 (3 + 4) × 3

7 (4 + 2) × 5

8 600 + 70 + 3

9 200 – 30

10 Subtract 20 from 80.

11 Subtract 3 tens from 40.

12 What is the sum of 150 and 46?

13 What is the product of 200 and 10?

14 How much are 3 magazines at $1.20 each?

15 

How much are 3 bottles of soft drink at $2.20 a bottle?

$ ☐

SET 2 Division strategies

1	2	3
2)864	4)488	6)964

4	5	6
4)886	3)690	4)840

7	8	9
3)972	6)684	6)786

10	11	12
4)968	3)762	7)868

13 Hanny won $756 in the lotto, which he had to share with another 3 people. How much did each person receive? 

14 George saved $616 in 8 months. What was his average saving per month? 

Space Measuring angles

1 Which angles measure 90°? ____________

2 Which angle measures 45°? ____________

3 Which angles measure 110°? ____________

4 Which angle measures more than 180°? ____________

5 Which angle measures 180°? ____________

B C D

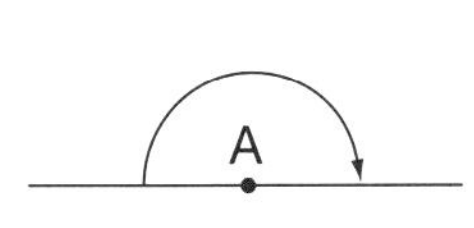

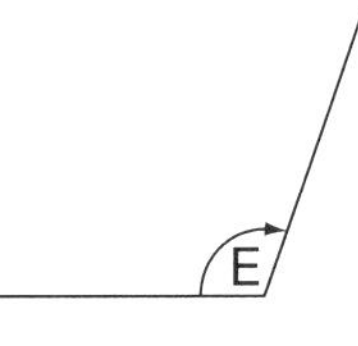

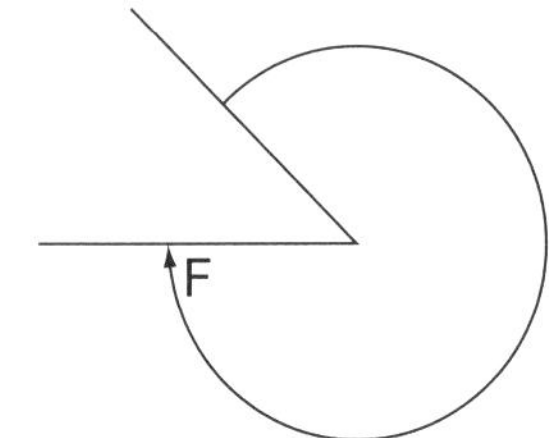

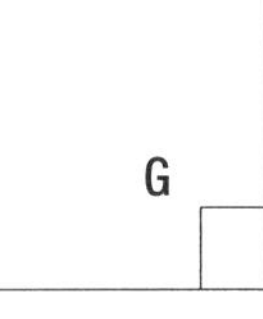

Number and Algebra

SET 3 Decimal tenths and hundredths

Write equivalent tenths, hundredths and decimals for each grid.

	Visual	Tenth	Hundredth	Decimal
1		$\frac{\quad}{10}$	$\frac{\quad}{100}$	.
2		___	___	.
3		___	___	.

4 Which is larger: $\frac{9}{10}$ or $\frac{93}{100}$?

5 Which is smaller: $\frac{3}{4}$ or 0.06?

6 0.76 equals ☐

7 Write 0.21 as a fraction.

8 Write $\frac{99}{100}$ as a decimal.

9 Write $\frac{87}{100}$ as a decimal.

10 Write $\frac{7}{10}$ as a decimal.

11 Write $\frac{1}{10}$ as a decimal.

SET 4 Extension

1 Share 40 among 6 people.

2 \$518 × 100

3 Double \$13.49

4 What is the area of a square with 5 cm sides?

5 What is the difference between 408 and 162?

6 \$8.78 × 8

7 $4\frac{1}{4}$ m = ☐ cm

8 50% of 80

9 Which is larger, 7^2 or $(7 \times 6) + 7$?

10 Write $\frac{7}{100}$ as a decimal.

11 27.6 – 4.3

12 If 3 cost 45c, how much would 7 cost?

13 If a square has a perimeter of 36 cm, what is the length of the base?

14 Write thirty-seven thousand, nine hundred and forty-two in figures.

Mathematical Reasoning

15 A palindromic number reads the same from right to left as it does from left to right, e.g., 3 1 5 1 3.

Make a palindromic number from these numbers: 2, 6, 3, 3, 2.

Measurement The tonne

Tick the box which best describes the mass of each item.

	Less than 1 t	About 1 t	More than 1 t
elephant			
25 children			
television			
car			
bulldozer			
refrigerator			

UNIT 17

Number and Algebra

SET 1 Basic

1 $\square \times 5 = 25$

2 $32 \div 8$

3 $20 \square 8 = 28$

4 $28 - 16$

5 $6^2 + 2$

6 $(3 \times 4) + 2$

7 $7 \square 5 = 35$

8 $(2 \times 6) + 4$

9 Divide 28 by 7.

10 $7.85 = ☐ c

11 115, 119, 123, ☐

12 What is the difference between 75 and 44?

13 Divide 12 by 12.

14 How much are 2 magazines at $2.10 each.

15

SET 2 Ordering decimals

Order the decimals from smallest to largest.

1	0.23	0.16	0.21	
2	0.73	0.52	0.17	
3	0.97	0.79	0.16	
4	0.17	0.01	0.27	
5	0.04	0.02	0.03	
6	1.27	2.71	1.07	
7	3.25	2.53	5.23	
8	7.21	1.27	2.12	
9	3.16	1.67	1.76	

Write the decimal that comes before and after the one supplied.

10		0.27	
11		1.27	
12		4.09	

Statistics and Probability Picture graphs

The following numbers were home runs hit by these baseball players.

Name	Home runs
Bart	𝍸 𝍸 ||
Betty	𝍸 𝍸
Homer	𝍸 𝍸 𝍸 |
Wilma	𝍸 𝍸 ||||
Luke	𝍸 𝍸

Create a pictograph from the home run data, using this key.

Key	(bat) = 2 runs

Bart Betty Homer Wilma Luke

Number and Algebra

SET 3 Using factors

Write all the factors for the following numbers.
The first one is done for you.

1	10	1, 2, 5 and 10
2	12	
3	15	
4	16	
5	20	
6	24	

Separate the second multiple into factors to find the answer to the following.

E.g. 42 × 12 becomes 42 × 4 × 3 = 504

7 23 × 12
8 31 × 15
9 41 × 12
10 52 × 15
11 43 × 14
12 39 × 12
13 40 × 20
14 32 × 16
15 44 × 16
16 33 × 20

SET 4 Extension

1 (49 ÷ 7) + 509
2 Average 30, 35, 27 and 40
3 50% of 460
4 29 × 100
5 Write these numbers in descending order 29 501, 48 602, 37 100.
6 What is the area of a rectangle with sides of 7 cm and 4 cm?
7 How many hundreds in 5280?
8 How many minutes from 7:45 am to 10:15 am?
9 20% of $45
10 $(6 \times 10^2) + (4 \times 8)$
11 Round 19 649 to the nearest thousand.
12 Share 97 between 6 people.
13 3.12, 3.18, 3.24, ☐ , ☐
14 Find the lowest common multiple of 9 and 5.
15 Write the numeral for a quarter of a million.
☐
16 If a rectangle has a perimeter of 90 cm and one side is 17 cm, what is the length of the sides that are not 17 cm?

Space Properties of prisms

Write a description of each object.

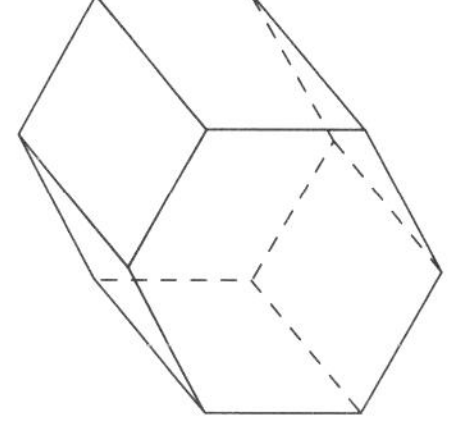

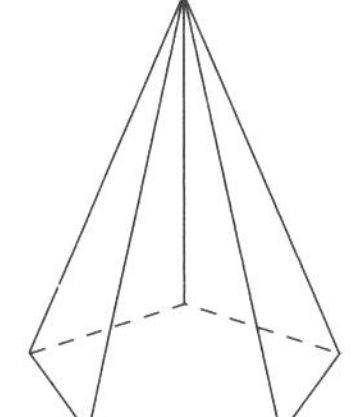

UNIT 18

Number and Algebra

SET 1 Basic

1 6 × 9

2 11 – 5

3 14 + 3

4 35 ÷ 7

5 28 ☐ 4 = 7

6 56 ☐ 8 = 64

7 60 ☐ 12 = 48

8 7 ☐ 6 = 42

9 213, 219, 225, ☐

10 180 minutes = ☐ hours

11 What is the value of 3 in 31 246?

12 How much are 3 rulers at 25c each?

13 Divide 40 by 5.

14 What is the product of 9 and 6?

15 Elle has 12 orange quarters cut up for the netball team. How many oranges did she cut up?

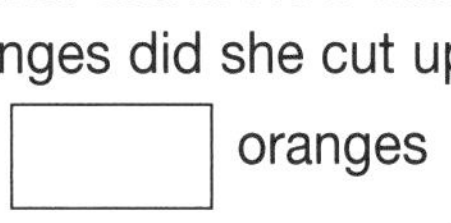

☐ oranges

SET 2 Rules of divisibility

- If the sum of the digits of a number is **divisible by 3** then the number is divisible by 3 as well.
- If the sum of the digits of a number is **divisible by 9** then the number is divisible by 9 as well.

Use this knowledge to answer true or false to the following statements.

1 15 is divisible by three.

2 33 is divisible by three.

3 59 is divisible by three.

4 78 is divisible by three.

5 127 is divisible by three.

6 453 is divisible by three.

7 99 is divisible by nine.

8 369 is divisible by nine.

9 56 is divisible by nine.

10 396 is divisible by nine.

11 785 is divisible by nine.

12 666 is divisible by nine.

Number and Algebra Financial plan/budget

Complete the balance column for the last 4 items – Petrol, Bills, Drinks and Movies.

	A	B	C	D
1	**Date**	**Item**	**Cost**	**Balance**
2	1 June	Opening		$400
3	7 June	Food	$160	(D2 – C3) $240
4	10 June	Petrol	$40	(D3 – C4)
5	11 June	Bills	$120	(D4 – C5)
6	20 June	Drinks	$30	(D5 – C6)
7	29 June	Movies	$20	(D6 – C7)

Number and Algebra

SET 3 Mixed numerals on a number line

Complete the number lines.

1

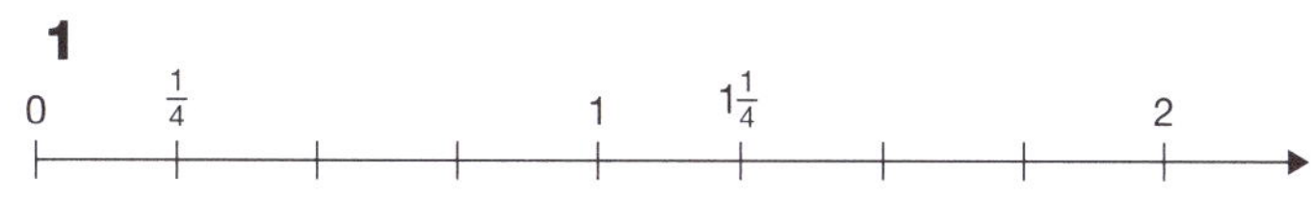

2

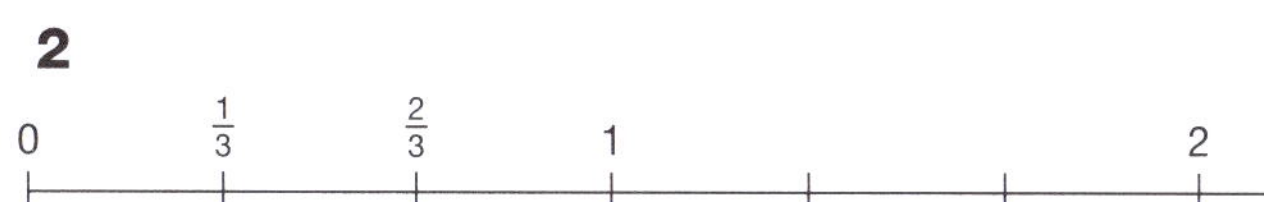

3

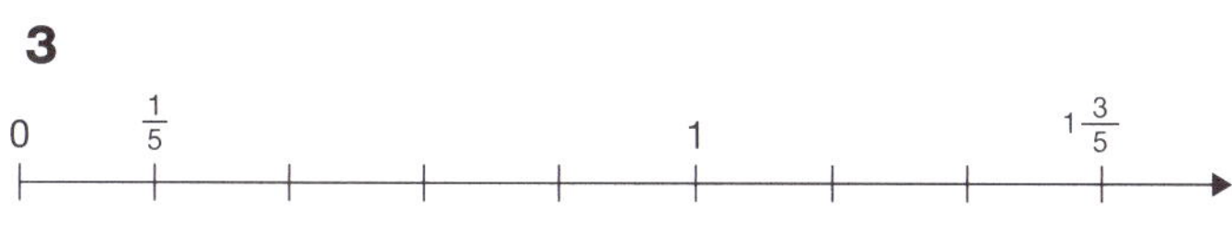

4

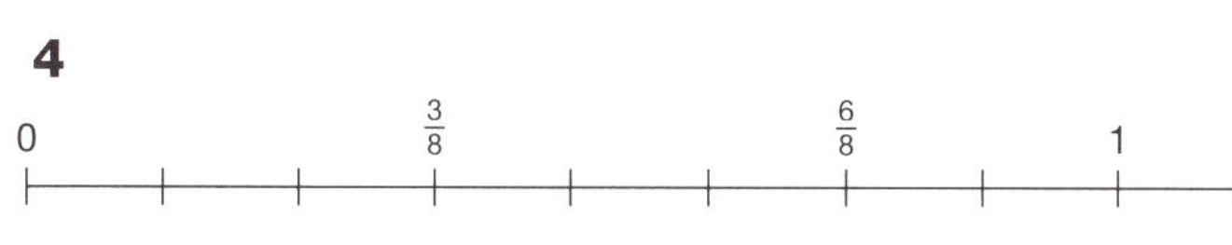

5

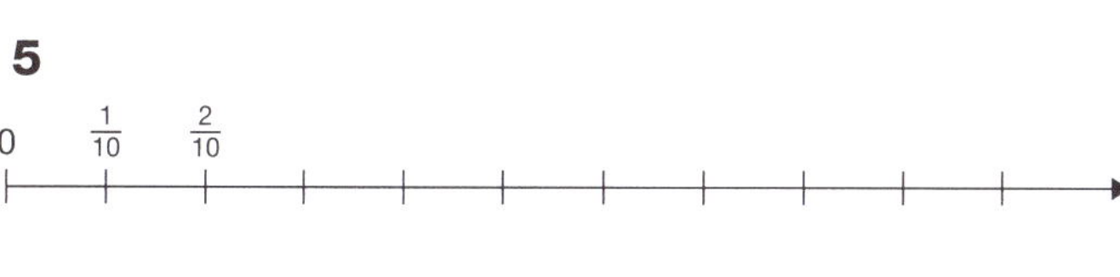

6

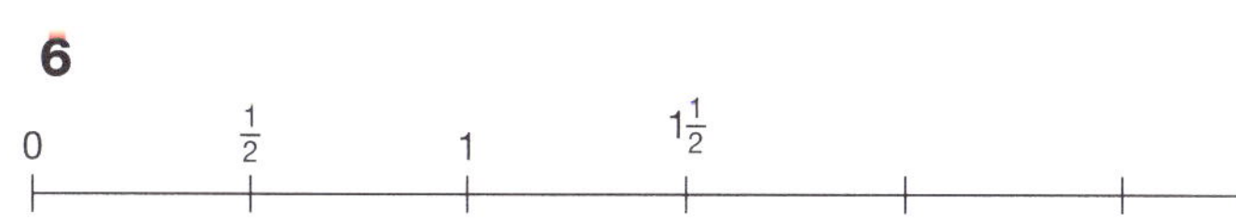

SET 4 Extension

1 $2^2 + 5^2$

2 79×1000

3 Write prime numbers between 6 and 20.

4 Write $4\frac{17}{100}$ as a decimal.

5 $\frac{4}{5}$ of 1 kg

6 How many minutes between 4:23 and 6:37?

7 Round 37 212 to the nearest hundred.

8 What is the perimeter of a hexagon with 27 cm sides?

9 25% of 60

10 $7^2 + 6$

11 $0.5 = \frac{}{10} = \frac{}{100} = \frac{}{2}$

12 If 5 cost 75c, how much would 9 cost?

13 How many seconds in $2\frac{1}{4}$ minutes?

14 Write these numbers in descending order 1479, 256, 35 256.

15 

If 1 mL of water has a mass of 1 g and volume of 1 cm^3, what would be the mass of a 2 L bottle?

Mathematical Reasoning

Statistics and Probability Databases

Netball database

	First	Surname	Address	Age	Grade
1	Jill	Arnott	Nerang	11	11A
2	Jess	Brown	Nerang	10	10B
3	Kelly	Cross	Southport	14	14C
4	Lauren	Ellis	Broadbeach	12	12A
5	Kate	Hood	Tugun	11	11A
6	Sophie	Jones	Carrara	11	11B
7	Eva	Kelly	Tugun	15	15C
8	Lena	Piper	Broadbeach	12	12A
9	Soula	Rhodes	Nerang	11	11A
10	Alisa	Smith	Tugun	16	16A
11	Auril	Suvan	Carrara	14	14B
12	Tim	Zatt	Tugun	10	10B

1 How many children live in Nerang?

2 How many children are 11 years old?

3 How many children live in Broadbeach?

4 How many children play in the 11A team?

5 How has this database been organised?

Number and Algebra

SET 1 Basic

1 14 + 6 + 5

2 19 – 5

3 6×8

4 $40 \div 8$

5 What is the product of 6 and 3?

6 What is the sum of 6, 4 and 9?

7 $4.15 = ☐ c

8 200 – 5

9 Double 12.

10 13 + 4 + 5

11 20 – 3

12 7×9

13 $42 \div 2$

14 What is the product of 7 and 4?

15 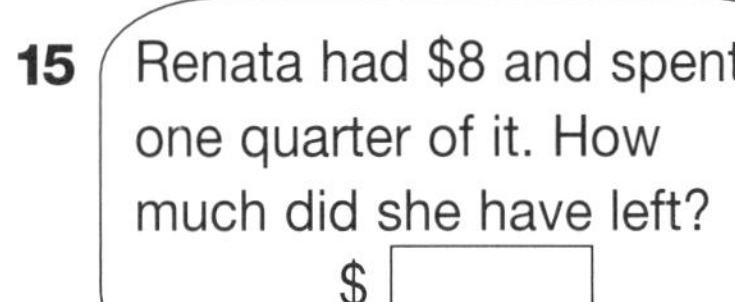

SET 2 Ordering 3-place decimals

Use the symbols $>$, $<$ or $=$ to make each number sentence true.

1 9.351 ☐ 9.513

2 6.264 ☐ 6.624

3 5.781 ☐ 5.788

4 3.254 ☐ 3.255

5 8.972 ☐ 8.9

6 2.199 ☐ 2.19

7 9.254 ☐ 9.452

8 6.36 ☐ 6.360

9 5.202 ☐ 5.2

10 4.434 ☐ 4.044

Write the decimal before and after these numbers.

11 ________ (9.515) ________

12 ________ (4.398) ________

13 ________ (6.799) ________

14 ________ (8.400) ________

15 ________ (3.601) ________

Number and Algebra Place value

1 Write each measurement on the decimal place value chart.

8.321 m 7.835 m 16.197 m 18.614 m 6.011 m

Metres					
10s of metres	Metres		$\frac{1}{10}$ metre	Centimetres	Millimetres

↓ ↓ ↓ ↓ ↓

Tens	Ones	•	Tenths	Hundredths	Thousandths
		•			
		•			
		•			
		•			
		•			

Number and Algebra

SET 3 Spreadsheets/fill down function

Follow the instructions below to find the multiples using Excel. You will need a computer with Microsoft Excel®.

1 Write the number 6 in cell A1.
Write the formula = A1+6 in cell A2.
Create the multiples by dragging down the cursor from A2 to A12.

2 Write the number 11 in cell B1.
Write the formula = B1+11 in cell B2.
Create the multiples by dragging down the cursor from B2 to B12.

3 Write the number 8.5 in cell C1.
Write the formula = C1+8.5 in cell C2.
Create the multiples by dragging down the cursor from C2 to C12.

4 Write the number 23 in cell D1.
Create a formula yourself and find the multiples.

5 Write the decimal 0.415 in the cell E1.
Create a formula yourself to find the multiples.

SET 4 Extension

1 How much are 5 books at $7.50 each?

2 Add 10 000 to 54 962.

3 $\frac{3}{5}$ of 70

4 ($\frac{3}{4}$ of 36) × 9

5 Which is larger, 112 or 10 × 10 × 10?

6 Round 10 961 to the nearest thousand.

7 Estimate an answer to 188 + 487.

8 Increase 9561 by four tens.

9 Average 12, 14, 18, 16

10 26.8 + 2.1

11 0.75 of a kilometre = ☐ m

12 (1000 × 8) + (100 × 6) + (10 × 5) + 9

13 How much is 4.5 kg at $6 per kg?

14 Which is larger, 3.5 or $3\frac{3}{10}$?

15 Write the numeral for (4 × 10 000) + (6 × 1000) + (3 × 100) + (9 × 10) + 8

Mathematical Reasoning

16 Jane had $456 in her purse but spent $45 on dinner, $126 on a new dress, $168 on an MP3 player and lost $50.

How much did she have left? ________

Space Triangles

What size are the missing angles?

1 Right angle triangle

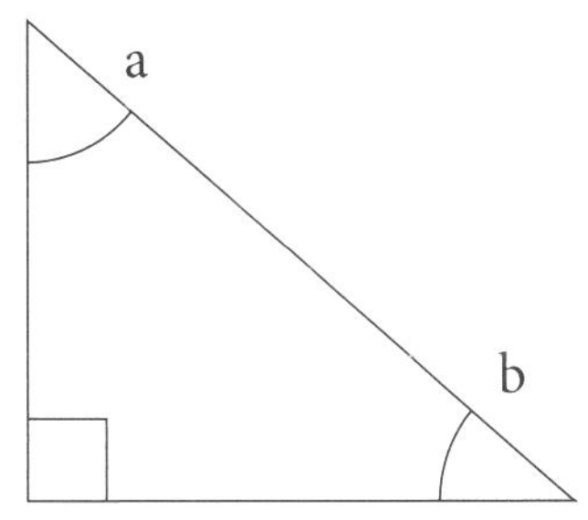

2 Isosceles triangle

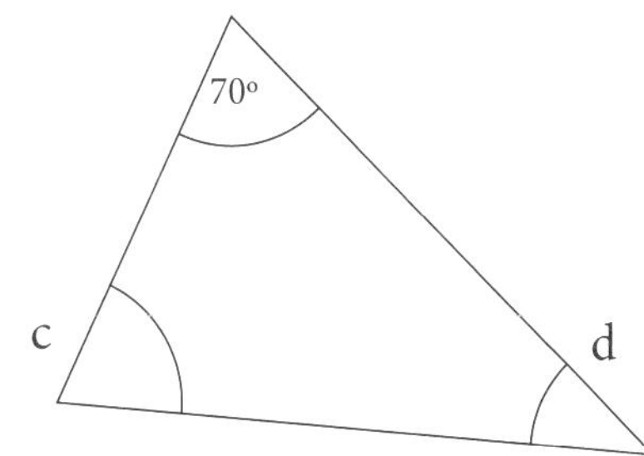

3 Scalene triangle

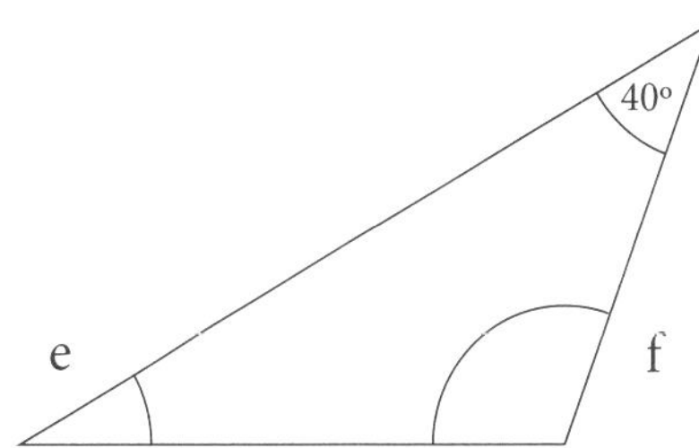

UNIT 20

Number and Algebra

SET 1 Basic

1 15 + 4 + 1

2 18 – 6

3 7 × 5

4 35 ÷ 7

5 What is the sum of 9 and 5?

6 $3.71 = [ ] c

7 150 – 6

8 Triple 8.

9 14 + 12 + 2

10 40 – 7

11 8 × 9

12 9^2

13 56 ÷ 8

14 Product of 8 and 6

15 Maria is 176 cm tall. If her sister, Soula is 32 cm shorter, how tall is Soula? [ ] cm

SET 2 Remainders as fractions

Complete the divisions.

1 4)496

2 3)486

3 5)585

4 5)680

5 6)721

6 5)682

7 4)375

8 3)551

9 6)683

10 7)891

11 8)860

12 7)771

13 8)983

14 6)673

15 5)871

16 7)953

Statistics and Probability Pie charts

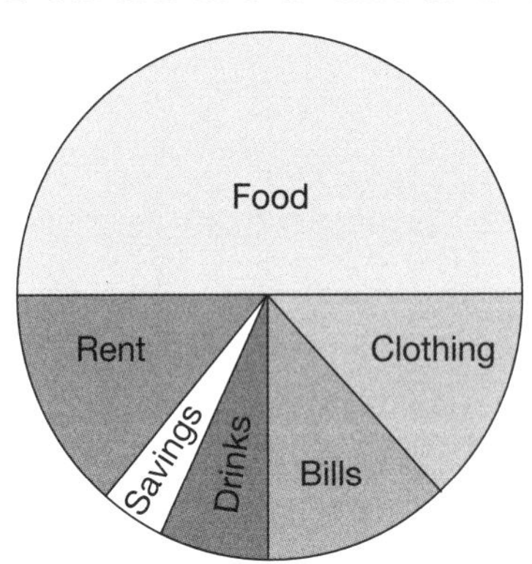

Anna has $400 per week to spend on her weekly expenses.

1 About how much do you think Anna spends on food?

2 About how much do you think she spends on bills?

3 Does Anna spend more on rent than drinks?

4 If Anna spends $30 on drinks, how much do you think she saves?

Number and Algebra

SET 3 Problem solving

Use your calculator's memory function to solve these problems.

1 Jim bought 5 computers for $1987 each. How much did he spend? $ ☐

2 Jessica bought 5 T-shirts for $29 each and 3 books at $17 each. How much did she spend? $ ☐

3 Thomas bought 2 model aeroplanes for $197 each and 6 packets of glue at $3.50. How much did he spend? $ ☐

4 Kim saved $137 a month for 6 months and $209 a month for another 6 months. How much did Kim save? $ ☐

5 Jenny wants a new TV set for $537. If she saves $47 a month for 7 months, how much more will she need? $ ☐

6 Peter bought a box of 48 apples for $16.80. How much was each apple? $ ☐

SET 4 Extension

1 2.6 + 3.3

2 Add 10 000 to 24 985.

3 $(4 \times 5) + (2 \times 10)$

4 How much is 4.5 kg at 80c per kg?

5 Which is larger, (4×10^2) or 4000?

6 Round 14 692 to the nearest ten thousand.

7 If 3 cost 66c, how much would 7 cost?

8 Decrease 7856 by four thousand.

9 Write a quarter to eleven in digital time.

10 Write ten thousand and six in numerals.

11 80% of a metre = ☐ cm

12 Which is larger, 7.23 or 3.99?

13 Round 6222 to the nearest hundred.

14 ($20 – 1) × 7

15 Write the numeral for $(5 \times 10\,000) + (7 \times 1000) + (4 \times 100) + (0 \times 10) + 9$

16 Write twenty-five past seven in the evening in 24-hour time.

Space Plan view

The school needs to order new floor covering for this classroom. Use the scale to calculate the square metres of carpet and lino needed.

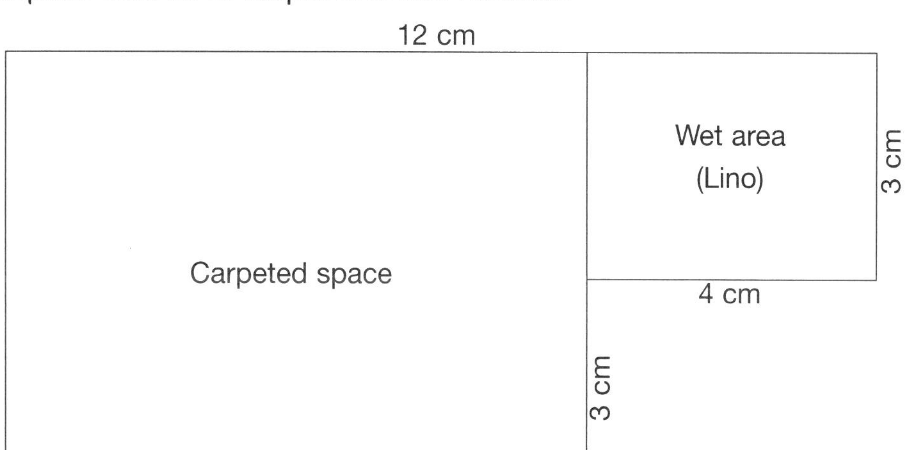

	Length	Width	Area
Carpet			
Lino			

Scale 1 cm = 1 m

Number and Algebra

SET 1 Basic

1 17 + 8

2 27 ÷ 3

3 17 + 24

4 8 × 10

5 26 – 15

6 8 × 4

7 13 ÷ 6

8 What is the value of 6 in 36 705?

9 18 ☐ 18 = 36

10 46 ☐ 12 = 34

11 What is the sum of 16 and 22?

12 What is the product of 6 and 8?

13 Write the Roman numeral for 27.

14 Share $4 among 4 people.

15

Carlos has 27 football cards but swapped 7 of them for 2. How many does he have now?

☐ cards

SET 2 Distributive law

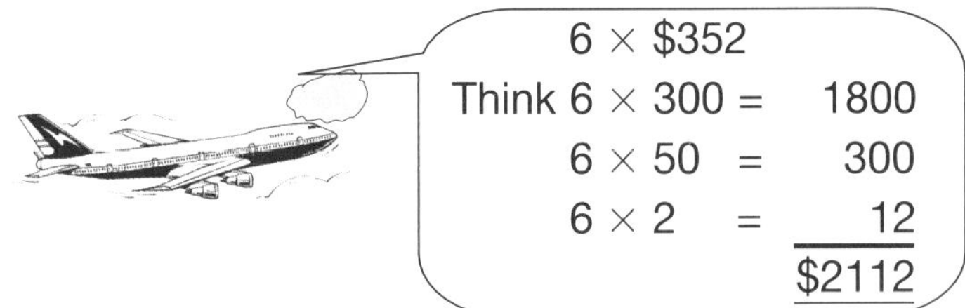

6 × $352

Think 6 × 300	=	1800
6 × 50	=	300
6 × 2	=	12
		$2112

	one-way	return
Melbourne – Cairns	$392	$516
Melbourne – Adelaide	$215	$343
Melbourne – Sydney	$258	$313

Use the distributive law to calculate the cost of the trips from Melbourne.

1 2 return tickets to Cairns

2 6 one-way tickets to Sydney

3 4 return tickets to Adelaide

4 8 one-way tickets to Adelaide

5 Would you save money by buying a return ticket if you were coming back?

6 Mr Cook bought 4 return tickets that cost him $1252. What was his destination? 

7 Mrs Hill makes 3 return trips to Sydney and 5 return trips to Adelaide each year. How much does she spend? 

Number and Algebra Using spreadsheets

Complete this computer-based spreadsheet. The C column has the calculations listed and has been widened to allow for your answers.

	A	B	C
1	**Date**	**Deposit**	**Subtotal**
2	22 Mar	$217.30	(= C2) $217.30
3	28 Mar	$133.40	(= C2 + B3)
4	4 Apr	$168.10	(= C3 + B4)
5	13 Apr	$321.70	(= C4 + B5)
6	19 Apr	$298.90	(= C5 + B6)

Number and Algebra

SET 3 Adding fractions

Complete the calculations.

1 $\frac{1}{4} + \frac{1}{4}$

2 $\frac{2}{4} + \frac{1}{4}$

3 $\frac{1}{3} + \frac{1}{3}$

4 $\frac{1}{5} + \frac{2}{5}$

5 $\frac{3}{5} + \frac{1}{5}$

6 $\frac{7}{10} + \frac{2}{10}$

7 $\frac{3}{12} + \frac{6}{12}$

8 $\frac{23}{100} + \frac{31}{100}$

9 $1\frac{1}{5} + \frac{1}{5}$

10 $2\frac{1}{4} + \frac{1}{4}$

11 $3\frac{4}{10} + \frac{3}{10}$

12 $4\frac{1}{3} + \frac{1}{3}$

13 $5\frac{3}{8} + \frac{2}{8}$

SET 4 Extension

1 $9\frac{1}{2}$ m = ☐ cm

2 Write $4\frac{7}{100}$ as a decimal.

3 How many grams in 4.5 kg?

4 7.84 – 3.32

5 Perimeter of a square with sides of 14 cm

6 3.5 kg at $4 per kilogram

7 $2.75 × 6

8 3.15 + 2.07 + 3.1

9 What are the side lengths of a square if its area is 81 m^2?

10 Does a rectangular prism have an apex?

11

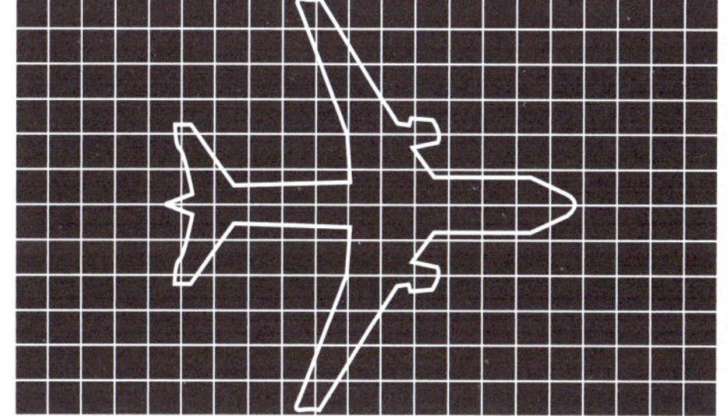

If each square on the grid measures 4 metres × 4 metres, estimate these lengths:

The length of the plane = ______ metres.

The width of the plane = ______ metres.

Mathematical Reasoning

Space Rotational symmetry

Tick the shapes that have rotational symmetry.

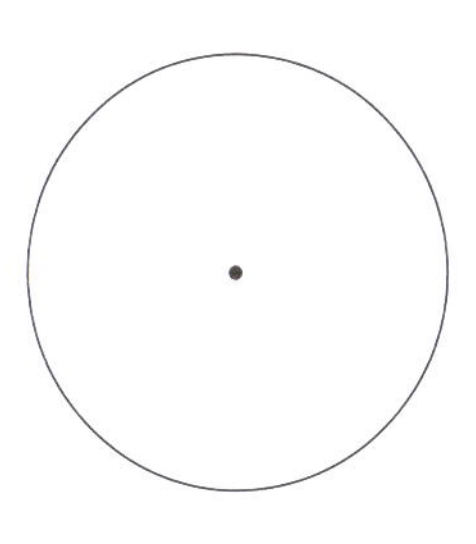

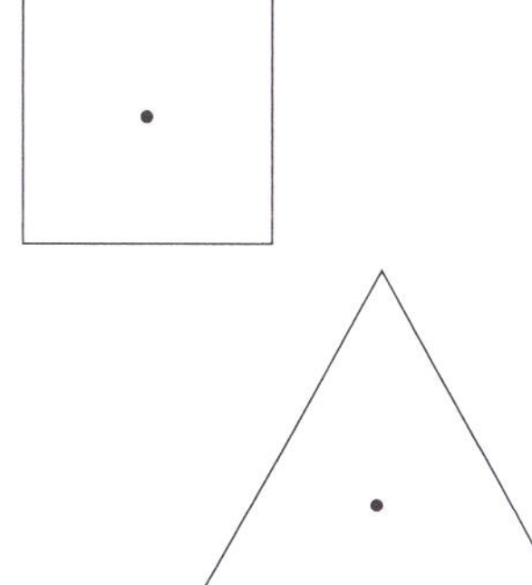

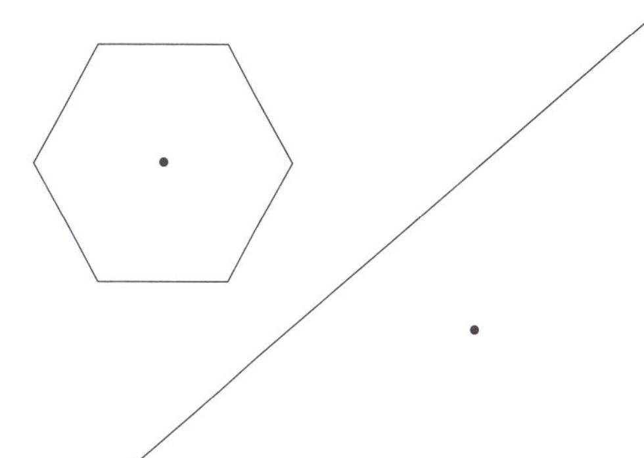

Number and Algebra

SET 1 Basic

1 7 ☐ 6 = 42

2 16 ☐ 4 = 20

3 What is the value of 1 in 10 374?

4 27 + 14

5 28 ÷ 7

6 17 + 7

7 9 × 6

8 8 × 9

9 43 – 8

10 24 ÷ 4

11 What is the difference between 30 and 17?

12 What is the 10th month of the year?

13 Share $54 among 6 people.

14 1248, 1240, 1232, ☐

15 Costa laid 24 bricks in 10 minutes. How many would he lay in 1 hour?

☐ bricks

SET 2 5-digit subtraction

1 $\$127.56 - 112.18$

2 $\$364.88 - 158.93$

3 $54\,800 - 37\,283$

4 $16\,395 - 7\,648$

5 $83\,754 - 55\,555$

6 $50\,907 - 49\,013$

Mathematical Reasoning

7 Which two towns am I thinking of if the difference in population between them is 18 444?

Town	Population
Blake	33 709
Rosemont	30 102
Linberg	100 468
Pinehurst	42 848
Greenfield	52 153
Lacey	30 705
Cotter	27 246

Statistics and Probability Line graphs

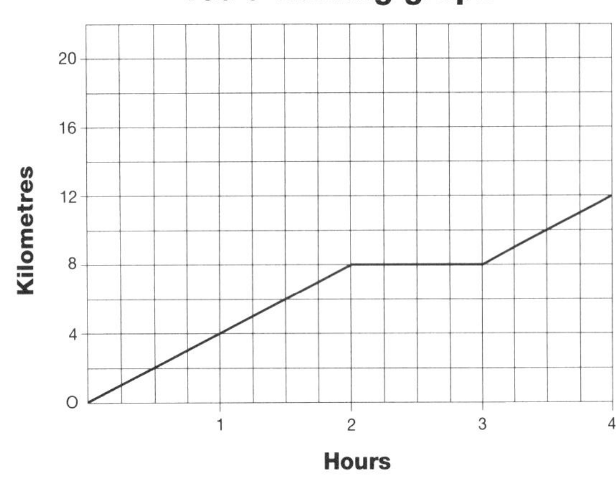

1 How far had Ted walked after 1 hour?

2 How far had Ted walked after 3 hours?

3 How far had Ted walked after 4 hours?

4 How far did Ted walk between the 2nd and 3rd hour?

5 What was Ted's average speed for the walk?

Number and Algebra

SET 3 GST

Add 10% ($\frac{1}{10}$) to the cost of an item.

Find the total cost of each manufactured item when the 10% GST is added.

1 

$40

GST	
Total cost	

2 

$90

GST	
Total cost	

3 

$60

GST	
Total cost	

4 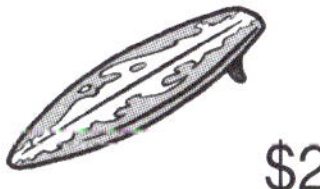

$200

GST	
Total cost	

5 

$160

GST	
Total cost	

SET 4 Extension

1 Do the hands on a clock at 5 pm form an obtuse angle?

2 Write $14\frac{9}{100}$ as a decimal.

3 0.8 m = ☐ cm

4 Round 3.79 to the nearest tenth.

5 What is the radius of a circle if 10 cm is the diameter?

6 How much is 8 kg of meat at $6.75 per kilogram?

7 Hundredths in 1.76

8 Perimeter of an octagon with 8 cm sides

9 Which is larger: 33%, $\frac{30}{100}$, or 0.34?

10 8.6 m = ☐ cm

11 What fraction of a century is 25 years?

12 $22 + 3^2 + 4^2$

13 $\frac{2}{3}$ of $36

14 How much is $3\frac{1}{2}$ m of timber at $7.40 per metre?

15 Average 92, 37, 101, 10

Measurement Timetables

Vegieville morning bus timetable

Carrot	Spud	Vegie	Pumpkin	Bean	Onion
5:55	5:58	6:02	6:10	6:20	
6:25	6:28	6:32	6:40	6:50	6:55
6:55	6:58	7:02	7:10	7:20	7:25
7:25	7:28	7:32	7:40	7:50	7:55
7:55	7:58	8:02	8:10	8:20	8:25
8:25	8:28	8:32	8:40	8:50	8:55
8:52	8:56	9:00	9:10	9:20	9:28
9:12	9:16	9:20	9:30	9:40	9:48
9:32	9:36	9:40	9:50	10:00	10:08
9:52	9:56	10:00	10:10	10:20	10:28
10:12	10:16	10:20	10:30	10:40	10:48
10:32	10:36	10:40	10:50	11:00	11:08
10:52	10:56	11:00	11:10	11:20	11:28
11:12	11:16	11:20	11:30	11:40	11:48
11:32	11:36	11:40	11:50	12:00	12:08
11:52	11:56	12:00	12:10	12:20	12:28

1 If you caught the 6:55 am from Carrot, when would you arrive at Onion? ______

2 If you caught the 7:02 am from Vegie, when would you arrive at Bean? ______

3 What bus would I need to catch from Spud to be at Onion by 10:48 am? ______

4 What bus would I need to catch from Vegie to be at Bean by 10:00 am? ______

5 How long does the 6:25 bus from Carrot take to reach Onion? ______

UNIT
23

Number and Algebra

SET 1 Basic

1 10×5

2 $40 + 8$

3 $21 - 3$

4 $42 \div 6$

5 6×9

6 $40 + 7$

7 $35 - 6$

8 $18 \div 9$

9 What is the product of 4 and 0?

10 3×7

11 $25 + 4$

12 $19 - 8$

13 $36 \div 4$

14 What is the sum of 8, 6 and 9?

15 Jade was given \$10, \$15, \$20 and \$25 for her birthday. How much did she receive altogether?

\$ 

SET 2 4-digit multiplication

Round the 4-digit numbers to the nearest thousand to estimate these products.

1 $4895 \times 5 =$

2 $7106 \times 9 =$

3 $5987 \times 6 =$

4 $8214 \times 8 =$

5 $6037 \times 3 =$

6 $9320 \times 2 =$

7 $6988 \times 7 =$

8 $5895 \times 4 =$

9 $8171 \times 5 =$

10 $7202 \times 8 =$

Solve these multiplications.

11 4567×6

12 3894×8

13 3579×7

14 6530×9

15 How much money was collected if 5 people each paid \$7654 for a trip?

Number and Algebra Number patterns with fractions

Complete the sequence, then write a rule for each one.

1

$\frac{1}{10}$	$\frac{2}{10}$	$\frac{3}{10}$		

2

$\frac{1}{12}$	$\frac{3}{12}$	$\frac{5}{12}$		

3

$\frac{1}{8}$	$\frac{3}{8}$	$\frac{5}{8}$		

4

$2\frac{1}{4}$	$3\frac{1}{4}$	$4\frac{1}{4}$		

5

$2\frac{1}{2}$	3	$3\frac{1}{2}$		

6

$1\frac{1}{4}$	$1\frac{2}{4}$	$1\frac{3}{4}$		

Number and Algebra

SET 3 Basic percentages

	Fraction	Decimal	%
1	$\frac{10}{100}$		
2	$\frac{25}{100}$		
3	$\frac{7}{10}$		
4	$\frac{20}{100}$		
5	$\frac{1}{2}$		
6	$\frac{1}{4}$		
7	$\frac{3}{4}$		

Order from smallest to largest.

8	$\frac{27}{100}$	30%	0.29	
9	35%	$\frac{53}{100}$	0.33	
10	$\frac{99}{100}$	9%	0.9	
11	0.54	$\frac{1}{2}$	49%	
12	4%	$\frac{3}{10}$	0.21	
13	$\frac{9}{100}$	90%	0.95	
14	$\frac{70}{100}$	7%	0.03	

Mathematical Reasoning

SET 4 Extension

1 $\frac{4}{5} + \frac{3}{5}$

2 What is the value of 7 in 37 829?

3 How many lines of symmetry has a rectangle?

4 4550 + 608

5 $(3 \times 10^2) + (4 \times 10) + 16$ ones.

6 How much are 8 books at $2.80 each?

7 $(\frac{3}{5} \times 50) + (\frac{3}{8} \times 40)$

8 If 5 cost 95c, how much would 40 cost?

9 How much are 4 combs at $4.50 each?

10 10:25 pm + 35 minutes

11 How many degrees in a quadrilateral?

12 19 000 – 200

13 Write one hundred and eighty-four in Hindu-Arabic numerals.

14 What is the perimeter of a rectangle 18 cm long and 6 cm wide?

15 Write twelve thousand, six hundred and twenty-nine in Hindu-Arabic numerals.

16 Bill's mass is 48 kg. If Sam's mass is the same as Bill's plus $\frac{1}{3}$, what is Sam's mass?

Statistics and Probability Chance

Shade the chances of the marbles being drawn out of the bag to match the fractions.

1 Red $\frac{4}{15}$

2 Blue $\frac{5}{15}$

3 Green $\frac{2}{15}$

4 Pink $\frac{1}{15}$

5 Yellow $\frac{3}{15}$

UNIT 24

Number and Algebra

SET 1 Basic

1 15 ÷ 3

2 18 + 18

3 32 + 14

4 7 × 6

5 43 – 6

6 4 × 8

7 8 × 4

8 36 ☐ 8 = 44

9 8 ☐ 7 = 56

10 What is the value of 6 in 37 364?

11 What is the sum of 17 and 24?

12 What is the product of 6 and 9?

13 Write the Roman numeral for 39.

14 13 750, 13 755, 13 760, ☐

15

Thelma and Louise saved $27 and $46. How much more do they need to save to buy a $85.50 present?

$ ☐

SET 2 Division strategies/ estimation

Round to the nearest 10 to estimate an answer for the following divisions.

1 68 ÷ 7 =

2 51 ÷ 10 =

3 89 ÷ 9 =

4 82 ÷ 10 =

5 148 ÷ 10 =

Solve these divisions.

6 8 ÷ 2 =

7 80 ÷ 2 =

8 800 ÷ 2 =

9 12 ÷ 4 =

10 120 ÷ 4 =

11 1200 ÷ 4 =

12 14 ÷ 7 =

13 140 ÷ 7 =

14 1400 ÷ 7 =

15 90 ÷ 9 =

16 900 ÷ 9 =

17 9000 ÷ 9 =

Measurement Perimeter

Create any two shapes with an area of 6 cm^2. One should have a perimeter of 10 cm and the other a perimeter of 14 cm.

Mathematical Reasoning

Number and Algebra

SET 3 Unknown quantities

Balance the equations by supplying the missing numbers.

1 $56 \div 8 = 28 \div \square$

2 $\square \times 9 = 42 + 30$

3 $96 \div \square = 4^2$

4 $120 \div 10 - \square = 29$

5 $\square \div 3 = 62 - 45$

6 $95 - 11 = \square \times 4$

7 $5^2 + 8 = 99 \div \square$

Mathematical Reasoning

Solve these missing number sentences. Each shape has the same value in every question, e.g., the triangle is always equal to 7.

8 □ + △(7) = 17

9 △ + ◯ = 19

10 ◯ + □ = ____

11 □ + ◯ + ⬡ = 25

12 ⬡ × ◯ + □ = ____

13 (□ − ⬡) × ◯ = ____

SET 4 Extension

1 How many centimetres in $9\frac{1}{4}$ metres?

2 $(15 + 8) + 5^2$

3 If 9 pencils cost 72c, how much would 17 cost?

4 How many degrees in a rectangle?

5 An obtuse angle is less than 90°. True or false?

6 3, 6, 12, 24, □

7 At 3, John is 75 cm tall. If he grows $5\frac{1}{2}$ cm a year, how tall will he be when he's 11?

8 I had \$73 but spent \$43.50. How much money have I now?

9 What is the perimeter of a hexagon with 15 cm sides?

10 7×836

11 13 750c = \$ □

12 Round off 31 326 to the nearest 10 000.

13 How many minutes from 10:43 am to 2:15 pm

14 Average 43, 27, 51, 13, 16

15 What is $\frac{4}{5}$ of \$90?

16 Write 10:45 pm in 24-hour time.

Space Coordinates

Join the coordinate sets to form a shape.

1 Join (B,1) to (B,7)

2 Join (B,7) to (F,7)

3 Join (F,7) to (F,6)

4 Join (F,6) to (D,6)

5 Join (D,6) to (D,5)

6 Join (D,5) to (F,5)

7 Join (F,5) to (F,4)

8 Join (F,4) to (D,4)

9 Join (D,4) to (D,1)

10 Join (D,1) to (B,1)

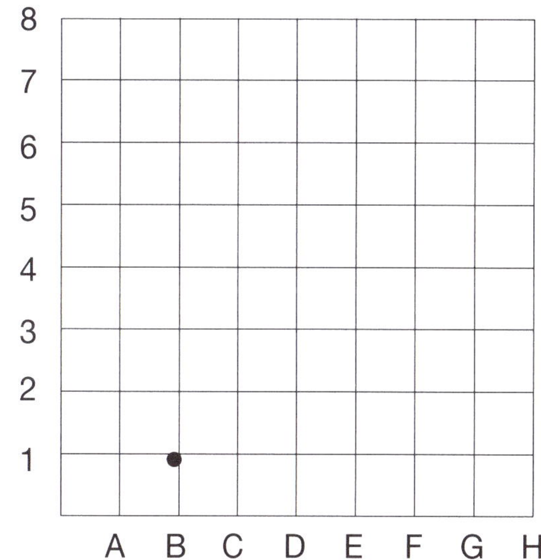

Number and Algebra

SET 1 Basic

1 18 + 3

2 17 − 13

3 13 + 16

4 9 × 4

5 16 − 12

6 9 × 6

7 45 ÷ 5

8 What is the value of 3 in 43 241?

9 8 ☐ 4 = 32

10 26 ☐ 17 = 9

11 What is the sum of 48 and 22?

12 What is the difference between 48 and 28?

13 How many days in August and September?

14 2 m = ☐ cm

15

Theo bought 7 boxes of bananas at $30 per box. How much did he spend?

$ ☐

SET 2 Multiplication by tens

	×	10	100	1000
1	2	20		
2	4			
3	8			
4	11			
5	12			

6 25×20

7 34×30

8 29×40

9 123×50

10 137×60

11 154×70

12 Mr King bought 139 computers at an auction for $70 each. How much did they cost him?

Measurement Perimeter and area

Calculate the perimeter and area of these shapes.

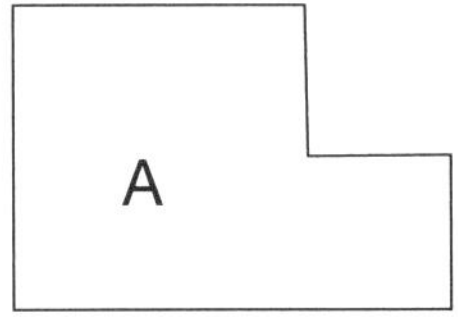

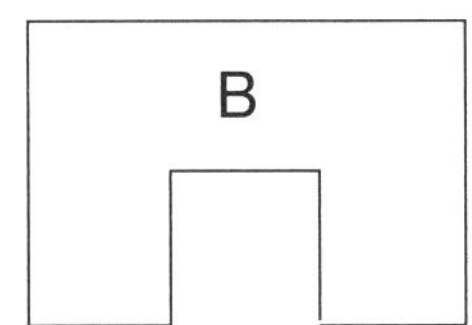

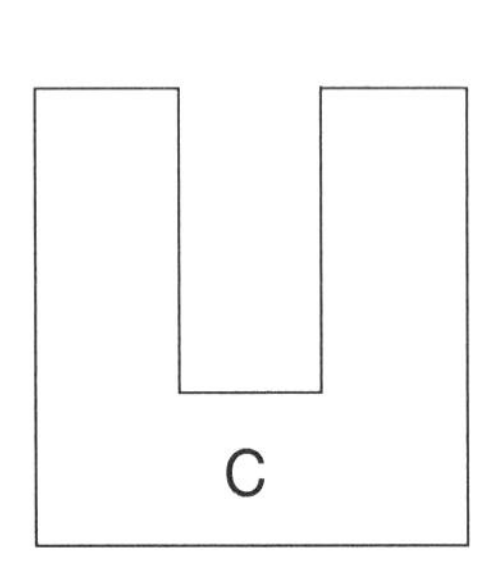

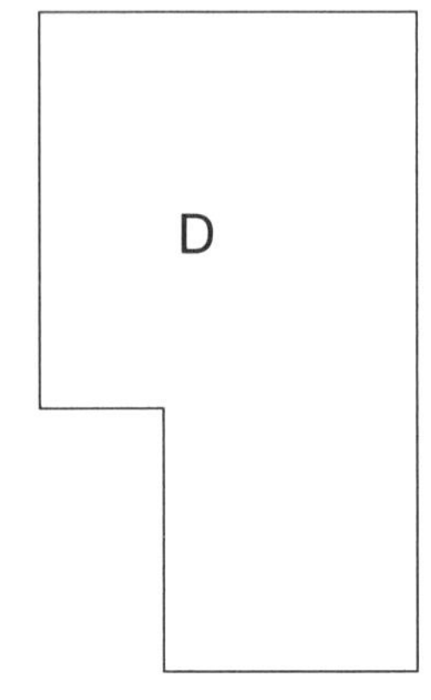

	Perimeter	Area
A	cm	cm^2
B	cm	cm^2
C	cm	cm^2
D	cm	cm^2

Number and Algebra

SET 3 Number patterns/ fractions/decimals

Continue the number patterns.

1	$\frac{1}{2}$	1	$1\frac{1}{2}$		
2	$\frac{1}{4}$	$\frac{2}{4}$	$\frac{3}{4}$		
3	$\frac{1}{3}$	$\frac{2}{3}$	1		
4	$\frac{1}{3}$	$1\frac{1}{3}$	$2\frac{1}{3}$		
5	2	$2\frac{1}{3}$	$2\frac{2}{3}$		
6	$1\frac{1}{8}$	$1\frac{3}{8}$	$1\frac{5}{8}$		
7	0.25	0.30	0.35		
8	0.36	0.44	0.52		
9	0.67	0.62	0.57		
10	0.05	0.10	0.15		

SET 4 Extension

1 0.3 of 1 m = ☐ cm

2 What is one third of 81?

3 An acute angle is less than 90°. True or false?

4 160, 80, 40, ☐

5 How many degrees between north and south?

6 Tim travelled for 6 hours at an average speed of 95 km/h. How far did he travel?

7 1350 + 4638

8 Round 3.69 to the nearest tenth.

9 How many eggs in $7\frac{1}{4}$ dozen?

10 The total mass of 6 men in a football scrum was 510 kg. What was the average mass?

Mathematical Reasoning

11 How many people play baseball, soccer and hockey if 50 play basketball and 10 play tennis?

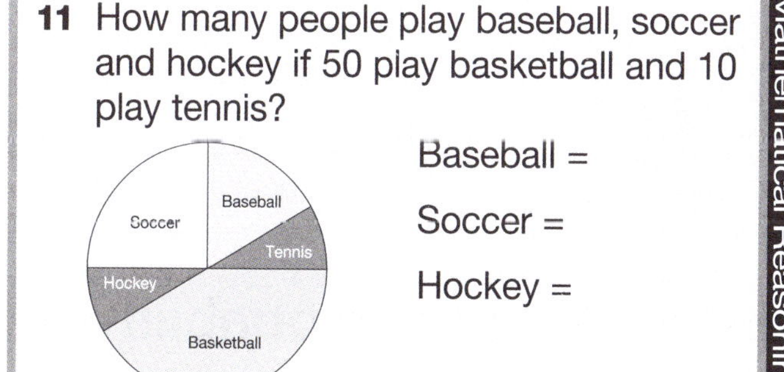

Baseball =

Soccer =

Hockey =

Space Drawing shapes

Construct a congruent copy of the shapes below. All angles and sides are given on the shapes.

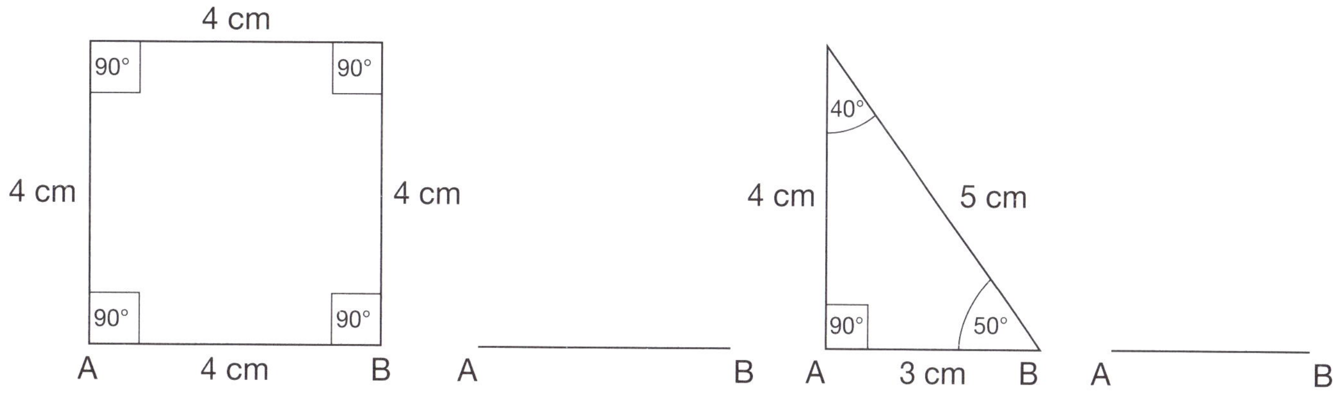

UNIT 26

Number and Algebra

SET 1 Basic

1 16 + 8

2 38 – 7

3 63 ÷ 7

4 What is the product of 8 and 7?

5 What is the sum of 9, 10 and 11?

6 4 + 4 + 6

7 27 – 8

8 20 ÷ 4

9 What is the average of 4, 6 and 5?

10 What is the difference between 21 and 3?

11 What is the cost of 4 books at $1.10 each?

12 400 cm = ☐ m

13 50 ÷ 10

14 60 – 40

15 Ricki puts 37 books on the library shelves every hour. If she works for 4 hours, how many books would she put on the shelves?

☐ books

SET 2 2-digit multiplication

1 33 × 16

2 31 × 53

3 18 × 75

4 27 × 33

5 82 × 21

6 45 × 42

7 12 × 27

8 31 × 22

9 62 × 99

10 There are 42 boxes of chocolates in the carton. Each box contains 24 chocolates. How many chocolates are there?

Statistics and Probability Data survey

Mr and Mrs King collected growth data on their son Sam.

1 Create a line graph from the information below. (Age is given in years and height in centimetres.)

Age	1	2	3	4	5	6	7
Height	60	70	80	90	100	110	120

Answer the questions.

2 How tall was Sam at 4 years?

3 How tall was Sam at $6\frac{1}{2}$ years?

4 How old was Sam when he was 100 cm tall?

Number and Algebra

SET 3 Equivalent number sentences

Circle the number that balances each equation.

	Equation	Possible solutions		
1	$15 \times 3 = \square + 8$	35	36	37
2	$10 + (\square \times 9) = 55$	5	10	15
3	$\square \div 9 \times 5 = 20$	45	36	54
4	$42 \div (3 + \square) = 6$	3	4	6
5	$81 - (\square \times 9) = 9$	8	9	72
6	$120 \div 10 \times \square = 12$	10	12	1
7	$54 \div 6 = \square \div 5$	44	45	89
8	$51 \div \square = 4^2 + 1$	17	3	13

Use = or ≠ to complete these equations.

9 $80 × 3 ☐ $60 × 4

10 1.5 kg × 3 ☐ 0.5 kg × 6

11 $0.90 × 6 ☐ $1.80 × 3

12 28 kg × 2 × 6 ☐ 9 kg × 5

13 2.5 m + 2.4 ☐ 0.7 m × 7

14 1500 mL × 500 ☐ 500 mL × 5

15 $21.90 ÷ 3 ☐ 73c × 10

SET 4 Extension

1 12 000 m = ☐ km

2 If a circle has a radius of 14 cm, what is its diameter?

3 How many thousands in 427 827?

4 What is the value of 6 in 5.06?

5 Round 16 295 to the nearest hundred.

6 If today is Tuesday 19 August, what is the date in 14 days' time?

7 $\square + 16 = 4^2$

8 A $250 bike is discounted by 10%. How much does it cost?

9 3.75 kg = ☐ g

10 75% of 200

11 30 L at $0.67 per litre

12 If 3 kg costs $54, how much would 8 kg cost?

Mathematical Reasoning

13 US$1.00 = A$0.98

1 American dollar is about equal to $0.98 Australian. What would be the value of US$9 in Australian dollars?

Measurement Grams and kilograms

How many of the smaller weights are needed to balance the large weight?

1

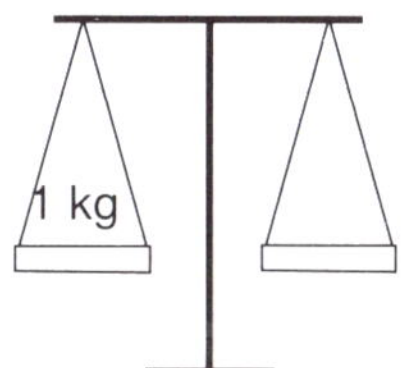

1 kg = 50 g × ☐

2

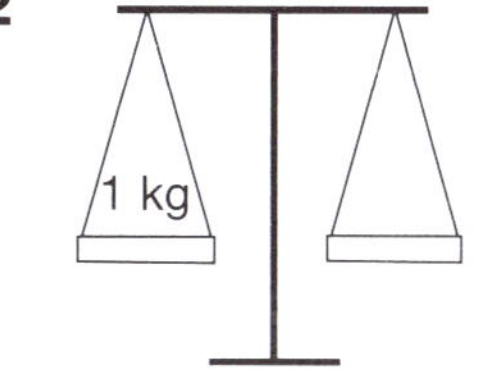

1 kg = 200 g × ☐

3

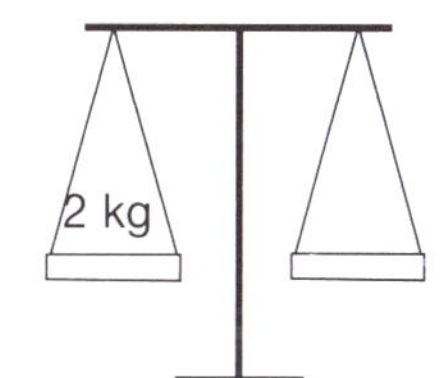

2 kg = 400 g × ☐

4

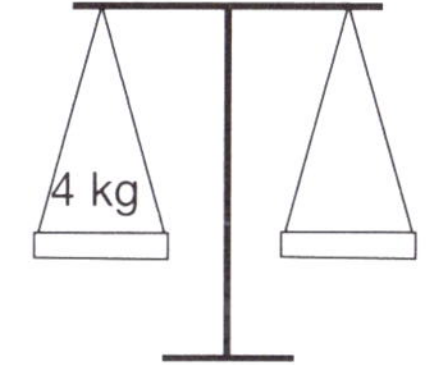

4 kg = 100 g × ☐

UNIT 27

Number and Algebra

SET 1 Basic

1 21 + 3 + 6

2 6 × 9

3 42 ÷ 7

4 What is the product of 7 and 4?

5 What is the sum of 9, 10 and 2?

6 $4.31 = ☐ c

7 What is the value of 4 in 34 329?

8 Share $16 among 4 people.

9 ☐ × 9 = 54

10 What is the cost of 2 tickets at $1.20 each?

11 18 × 10

12 (2 × 4) + 5

13 44 – 8

14 How many months are there in each season?

15 

Mr Hill made $375 last week and $504 this week. How much did he make in the fortnight?

$ ☐

SET 2 Division strategies

1 $\frac{840}{7}$

2 618 ÷ 6

3 $4\overline{)834}$

4 $\frac{832}{6}$

5 525 ÷ 5

6 $3\overline{)907}$

7 $\frac{630}{6}$

8 936 ÷ 9

9 $8\overline{)700}$

10 $\frac{700}{7}$

11 780 ÷ 10

12 $10\overline{)960}$

13 Our house has an area of 238 m^2. If there are 7 rooms, what is the average size of each room? 

14 Mohamed's group was given 840 blocks to share among their group of 8. How many did each person get? 

15 $606 was shared between 6 winners. How much did each person receive? 

Measurement Litres and millilitres

1 How many millilitres in 2 L? ________ mL

2 How many millilitres in 7 L? ________ mL

3 How many millilitres in $2\frac{1}{2}$ L? ________ mL

4 How many millilitres in $2\frac{1}{4}$ L? ________ mL

5 How many millilitres in $3\frac{1}{4}$ L? ________ mL

6 How many millilitres in $7\frac{1}{10}$ L? ________ mL

7 How many millilitres in $9\frac{3}{10}$ L? ________ mL

Number and Algebra

SET 3 Calculator division/ multiplication

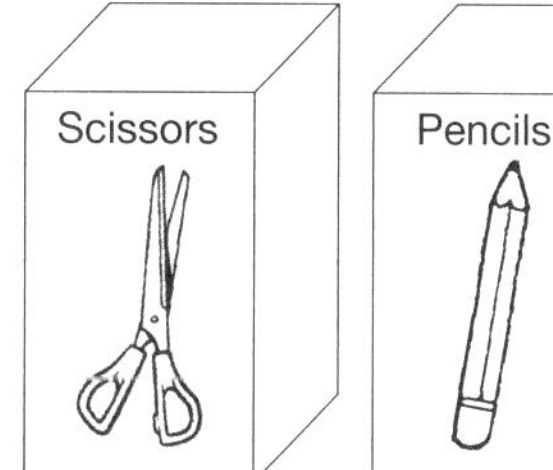

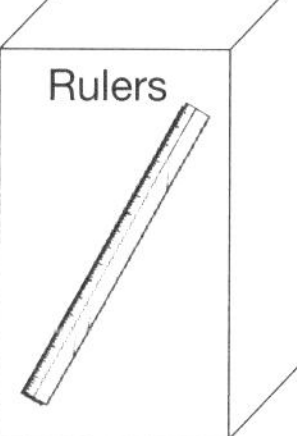

Pack of 20 $149.5	Pack of 34 $40.80	Pack of 25 $91.25	Pack of 72 $61.20

Calculate the cost of the individual items.

		Cost per item
1	Cost of a pair of scissors	$
2	Cost of a pencil	$
3	Cost of a glue stick	$
4	Cost of a ruler	$

Calculate the total cost for each product.

	Product	Price	Quantity	Total cost
5	Soccer boots	$74	46	$
6	Soccer ball	$59	72	$

SET 4 Extension

1 How much is 3.5 kg at $7 per kg?

2 Write 20 past 9 in digital form.

3 Round 17 653 to the nearest hundred.

4 60% of a metre = ☐ cm

5 Which is larger: (5×10^2) or 5000?

6 Write $15\frac{8}{10}$ as a decimal.

7 18 cm + 4 mm = ☐ mm

8 Hundredths in 4.64

9 How many 1.6 m lengths can I cut from an 8 m piece?

10 A cup holds 250 mL. How many would be needed to fill a $4\frac{1}{2}$ L jug?

11 How much is $4\frac{1}{2}$ m of ribbon at $3.20 a metre?

12 What is the average of 41, 49, 44 and 46?

13 Show all the possible totals when throwing two dice.

Dice	1	2	3	4	5	6
1	2	3	4	5	6	7
2	3	4	5	6		
3	4	5	6			
4	5	6				
5	6					
6						

Which total is the most common?

Mathematical Reasoning

Measurement Timelines

Draw a line to the timeline to place these events in Australian history.

Federation of Australia 1901

End of World War I 1918

End of World War II 1945

1900 — 1910 — 1920 — 1930 — 1940 — 1950 — 1960

Beginning of World War I 1914

Beginning of World War II 1939

Olympic Games in Melbourne 1956

Number and Algebra

SET 1 Basic

1 $5^2 + 3$

2 \$2.50 + 40c

3 3 m × 4

4 List the factors of 16.

5 Two places after 62nd

6 9 tens + 4 ones

7 How much is 6 m of material at \$20 per metre?

8 Divide 21 by 7.

9 95 × 0

10 \$4.10 + \$2.40

11 \$5.48 = ☐ c

12 8 × 0

13 What is half of 68?

14 72 = ☐ tens + ☐ ones

15

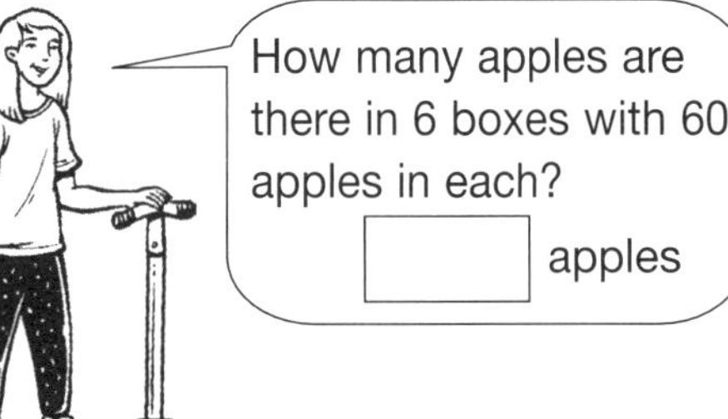

How many apples are there in 6 boxes with 60 apples in each?

☐ apples

SET 2 Backtracking/unknown quantities

What number did I start with?

1 ☐ multiply by 4, add 8 = 32

2 ☐ subtract 16, divide by 4 = 6

3 ☐ add 11, divide by 3 = 12

4 ☐ multiply by 10, subtract 8 = 62

5 ☐ divide by 8, add 19 = 24

6 ☐ subtract 9, divide by 5 = 9

7 ☐ add 13, multiply by 4 = 400

8 ☐ divide by 4, add 1 = 251

9 ☐ multiply by 8, subtract 10 = 62

10 ☐ subtract 14, add 19 = 43

11 How much money did Sally begin with if she brought home \$2, after paying \$4.50 on the bus, spending \$4.00 on snacks and \$9.50 on movie tickets?

12 How much water was in Tom's jug in the first place if he ended up filling the 1500 mL jug after adding 450 mL of crushed ice and 350 mL of cordial?

Measurement Cubic centimetres and cubic metres

Match the labels to the objects by drawing a line.

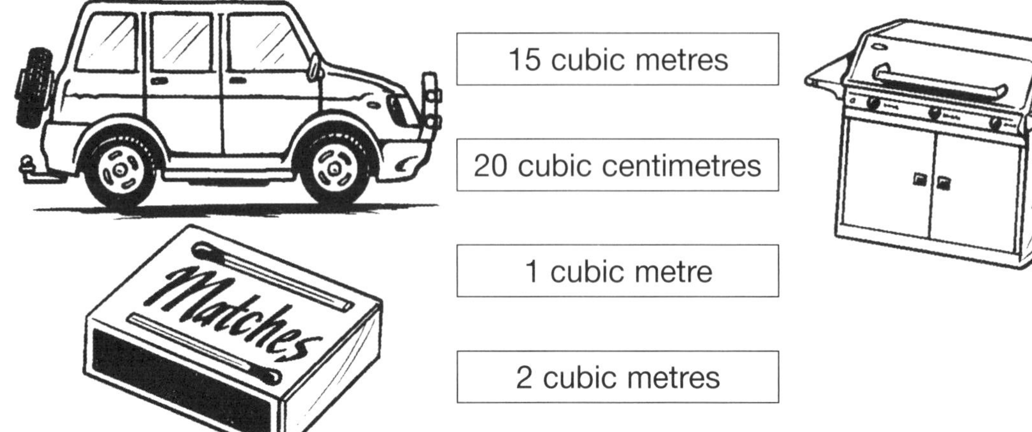

Number and Algebra

SET 3 Equivalence on number lines

1 Fill in the blank boxes to find the equivalent quarter or mixed numeral.

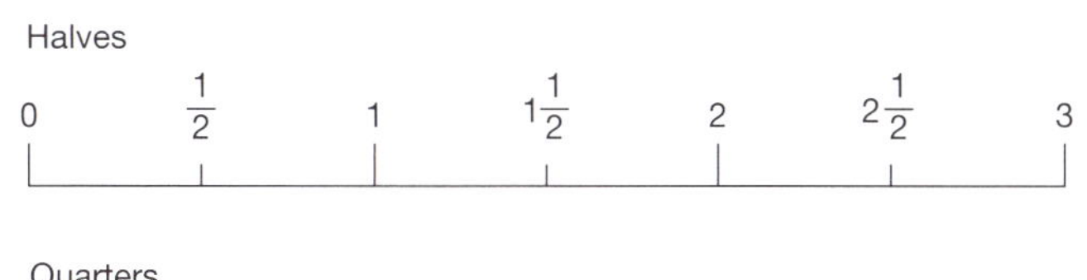

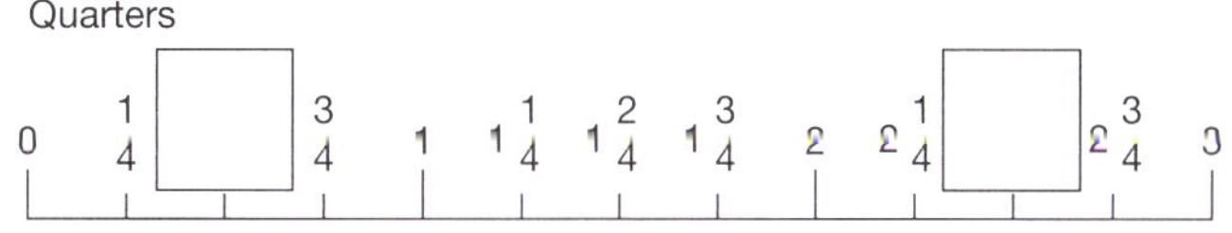

2 Fill in the blank boxes to find the equivalent fifth or mixed numeral.

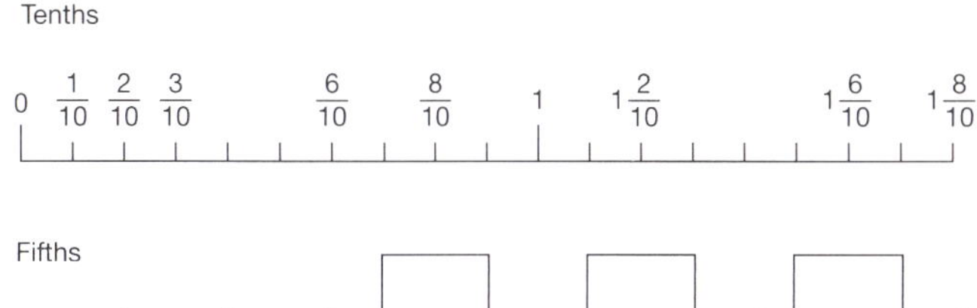

3 Fill in the blank boxes to find the equivalent quarter or mixed numeral.

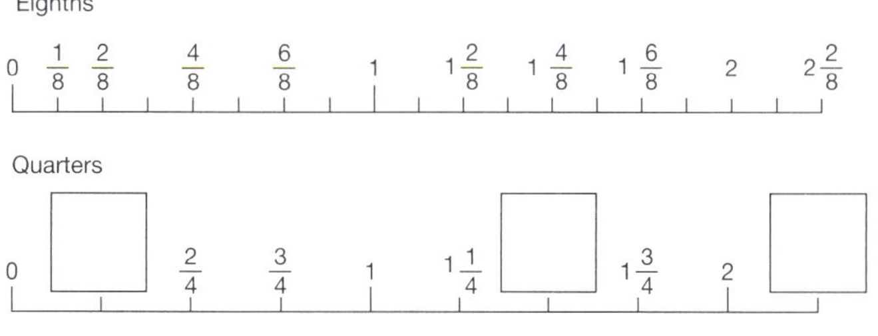

SET 4 Extension

1 What is the value of 6 in 3.61?

2 What is the perimeter of a 3.5 cm wide square?

3 Share $25 among 4 people.

4 6.51 pm + 18 min

5 Write sixty-four thousand, nine hundred and eight-two in Hindu-Arabic numerals.

6 What direction is 135° clockwise from North?

7 How much is 250 g of mince meat at $3.80 per kg?

8 If today is 27 September, what will be the date one fortnight from now?

9 4608 + 1300

10 $6.85 + $3.05

11 How many days in 3 years including a leap year?

12 A plane flew at 560 km/h for 3 hours. How far did it travel? 

13 If Chan has a one-fifth share in a boat valued at $655, what is the value of his share? 

14 How many degrees from north-east to north-west on a compass?

Statistics and Probability Line graphs

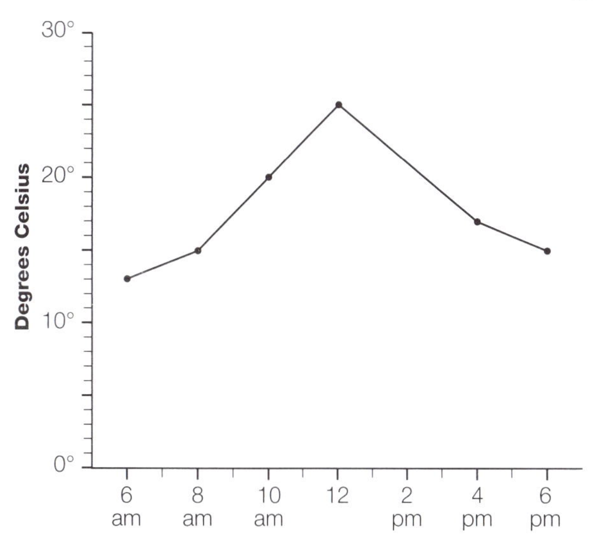

1 What was the temperature at 10 am? °C

2 What was the temperature at 4 pm? °C

3 How much did the temperature rise between 8 am and 12? °C

4 What is the difference between the highest and lowest temperatures? °C

5 Estimate the temperature at 2 pm. °C

UNIT 29

Number and Algebra

SET 1 Basic

1 20 ÷ 4

2 34 + 16

3 17 − 9

4 8 × 9

5 43 − 16

6 37 + 37

7 8 × 4

8 16 ÷ 3

9 What is the value of 6 in 15 620?

10 16 ☐ 4 = 4

11 26 ☐ 24 = 50

12 How much is half of $46?

13 Triple 9.

14 120 minutes = ☐ hours

15 If today is Monday the 8th, what day will it be on the 18th of the same month? ☐

SET 2 Adding and subtracting fractions

1 $\frac{7}{10} + \frac{2}{10} =$

2 $\frac{8}{10} - \frac{5}{10} =$

3 $\frac{9}{10} - \frac{4}{10} =$

4 $\frac{7}{8} - \frac{3}{8} =$

5 $\frac{4}{8} + \frac{1}{8} =$

6 $\frac{4}{5} - \frac{1}{5} =$

7 $\frac{3}{5} + \frac{1}{5} =$

8 $1 - \frac{3}{5} =$

9 $\frac{2}{6} + \frac{3}{6} =$

10 $\frac{7}{12} - \frac{5}{12} =$

11 $\frac{1}{6} + \frac{4}{6} =$

12 $\frac{3}{12} + \frac{8}{12} =$

13 $\frac{2}{6} + \frac{2}{6} =$

14 $\frac{3}{8} + \frac{4}{8} =$

Mamma's slab pizza

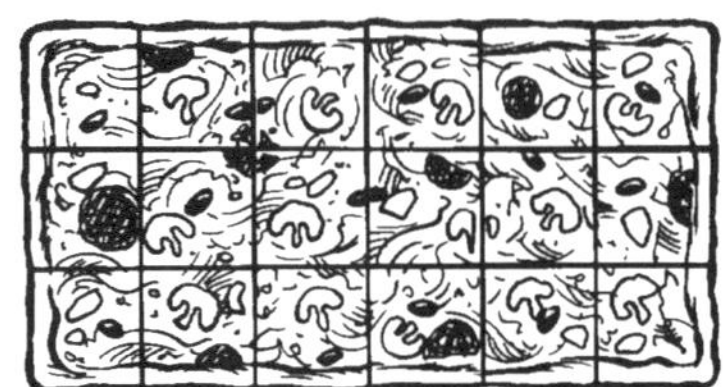

15 If Tom ate $\frac{3}{18}$, Jerome $\frac{5}{18}$ and Sally $\frac{4}{18}$, how much would be left?

Mathematical Reasoning

16 How could an 18-piece pizza be shared between two people so that one person receives twice as much as the other?

Number and Algebra Rounding for estimation

Round these numbers to the nearest 1000 in order to estimate the products.

	Number	x 4	x 5	x 10
1	1089	4000	5000	10 000
2	2189			
3	3588			
4	4768			
5	5217			
6	6891			

Mathematical Reasoning

Round each starting number to the nearest 10 or 100 to give quick estimations of the answers to these divisions.

	Number	Rounded to	Divided by	Estimate	Answer
7	397	400	4	100	99 r 1
8	789		8		
9	791		4		
10	149		5		
11	254		5		

Number and Algebra

SET 3 Percentages

Calculate the given percent of each number.

1 10% of 100
2 30% of 100
3 50% of 80
4 40% of 100
5 20% of 50
6 10% of 80
7 10% of 70
8 70% of 100
9 25% of 40
10 25% of 100

11 Shade 25% of the circles red.
12 Shade 15% of the circles blue.
13 Shade 20% of the circles yellow.
14 Shade 40% of the circles green.

SET 4 Extension

1 Write $27\frac{43}{100}$ as a decimal.
2 If 8 cost $56, how much would 7 cost?
3 I had $100 but spent $43. How much have I now?
4 What is the perimeter of an equilateral triangle with sides of 13 cm?
5 If there are 180 legs and there are 40 dogs, how many people are there?
6 Round 13.71 to the nearest tenth.
7 Average 27, 43, 30 and 20.
8 What fraction of 2 kg is 250 g?
9 How much is 8.5 kg of bacon at $5 per kilogram?
10 How many 125 mL containers would be needed to fill a container of 2.5 L?

Mathematical Reasoning

11 If a 2.5 metre signpost has a shadow 5 metres long, how tall is the flagpole if its shadow is 8 metres long?

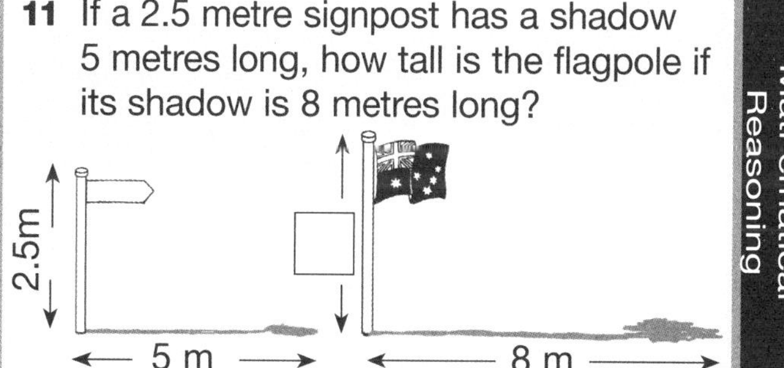

Measurement Tidal charts

Study the tidal chart then answer the questions.

Jan	QLD Brisbane				NSW Sydney (Fort Denison)			
	am	m	pm	m	am	m	pm	m
Fri 1st	2.43	1.73	2.36	1.84	1.48	1.33	1.36	1.27
	8.30	0.88	9.21	0.61	7.42	0.71	8.02	0.54
Sat 2nd	3.52	1.77	3.35	1.74	2.42	1.37	2.40	1.20
	9.45	0.94	10.21	0.59	8.51	0.72	8.55	0.57
Sun 3rd	5.03	1.87	4.45	1.68	3.36	1.42	3.48	1.17
	11.05	0.92	11.22	0.54	10.07	0.69	9.51	0.58
Mon 4th	6.06	2.02	12.19	0.84	4.30	1.50	4.54	1.17
			5.51	1.68	11.14	0.61	10.45	0.56
Tue 5th	12.18	0.46	1.23	0.74	5.20	1.59	12.09	0.51
	7.00	2.18	6.52	1.71			5.54	1.21
							11.36	0.52
Wed 6th	1.11	0.37	2.20	0.63	6.10	1.70	12.58	0.39
	7.50	2.34	7.48	1.76			6.47	1.26
Thu 7th	2.00	0.29	3.13	0.53	12.25	0.46	1.44	0.28
	8.37	2.48	8.40	1.82	6.59	1.81	7.37	1.32

1 Over how many days are the measurements recorded?
2 Where in Sydney are the tides measured?
3 How many measurements were usually recorded each day in Brisbane?
4 Give the highest tide measurement taken in Sydney over this period.
5 The lowest measurement taken in Brisbane was 0.29 m. On what date was this taken?

UNIT 30

Number and Algebra

SET 1 Basic

1 43 + 14

2 58 – 27

3 18 + 22

4 16 – 8

5 7 × 8

6 49 ÷ 7

7 64 ÷ 8

8 18 ☐ 3 = 6

9 17 ☐ 18 = 35

10 What is the product of 7 and 5?

11 Divide 36 by 6.

12 How many 5c coins in $1.15?

13 13 741, 13 751, 13 761, ☐

14 Write 95 in words.

15 

SET 2 Adding related fractions

1 whole							
			$\frac{1}{2}$				$\frac{2}{2}$
	$\frac{1}{4}$		$\frac{2}{4}$		$\frac{3}{4}$		$\frac{4}{4}$
$\frac{1}{8}$	$\frac{2}{8}$	$\frac{3}{8}$	$\frac{4}{8}$	$\frac{5}{8}$	$\frac{6}{8}$	$\frac{7}{8}$	$\frac{8}{8}$

Add the fractions.

1 $\frac{3}{8} + \frac{2}{8} =$

2 $\frac{4}{8} + \frac{1}{8} =$

3 $\frac{3}{4} + \frac{1}{4} =$

4 $\frac{1}{2} + \frac{1}{4} =$

5 $\frac{1}{2} + \frac{1}{8} =$

6 $\frac{1}{4} + \frac{1}{8} =$

7 $\frac{3}{4} + \frac{1}{8} =$

8 $\frac{3}{8} + \frac{1}{4} =$

9 $\frac{3}{8} + \frac{1}{2} =$

10 $\frac{5}{8} + \frac{5}{8} =$

11 $\frac{3}{4} + \frac{3}{4} =$

12 $\frac{7}{8} + \frac{7}{8} =$

Space Nets

Colour the nets that will fold to make a cube.

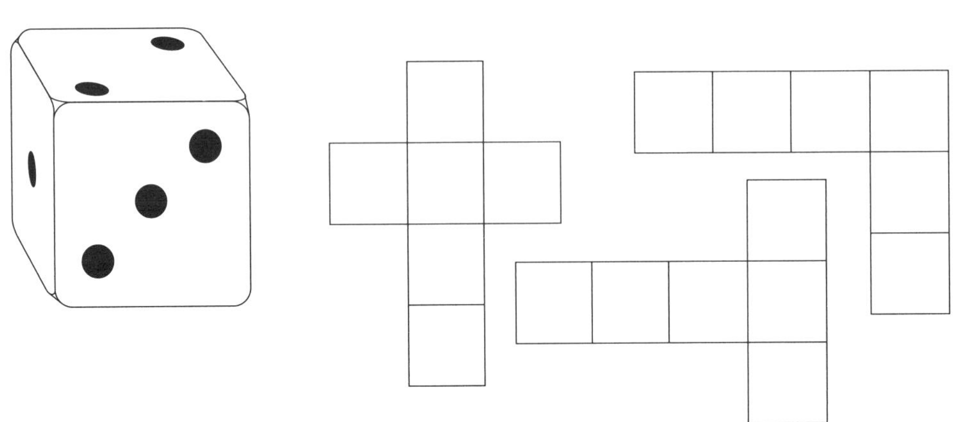

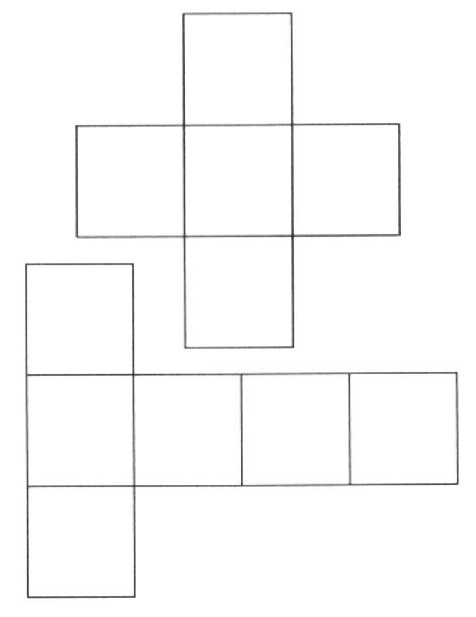

Number and Algebra

SET 3 Problem solving

Lucy's Boat Hire

Hours	1	2	4	6
Cost	$25	$48	$88	$120

Calculate these hire rates per hour.

1 2 hours

2 4 hours

3 6 hours

4 What did you notice about the cost per hour as the period of time got longer?

Calculate how much money Lucy received if she hired out these boats during the week.

		1 hr	2 hr	4 hr	6 hr	Total $
5	Mon	|||	|	|		$211
6	Tues	||		|		
7	Wed		|		|	
8	Thurs	|	|	|		
9	Fri	|	|||			

Mathematical Reasoning

SET 4 Extension

1 Area of a square with sides of 4 cm

2 How many 400 g packets in 2 kg?

3 $\frac{1}{4}$ of 2 km

4 1110, 1133, 1156, ☐

5 Value of 7 in 173.59

6 What fraction of $36 is $9?

7 Round 37 411 to the nearest hundred.

8 What is the perimeter of a decagon with sides of 17 cm?

9 $8^2 - 7$

10 $7\frac{1}{4}$ kg of sugar at $1.60 per kilogram

11 Write $17\frac{17}{100}$ as a decimal.

12 3 books at $7.20 each.

13 1, 3, 9, 27, ☐

14 Months in $4\frac{1}{2}$ years

15 What angle is formed by the hands of a clock at 6 o'clock?

16 If 7 cost $42, how much would $7\frac{1}{2}$ cost?

17 $60 × $80

18 Round off to the nearest $10 to answer $157.60 + $83.20.

Space Triangles and parallelograms

Draw a line to match the shapes to their names.

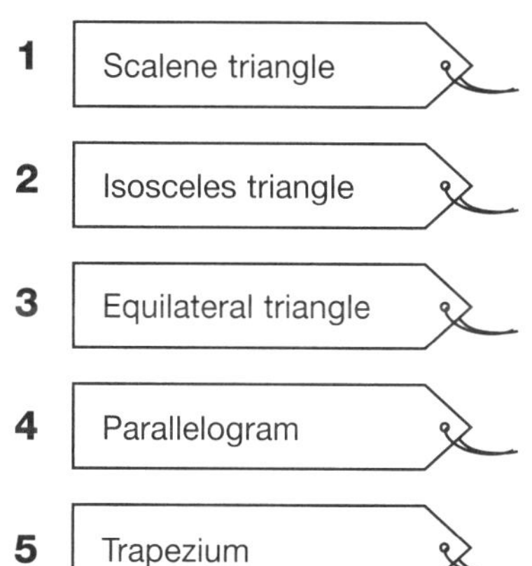

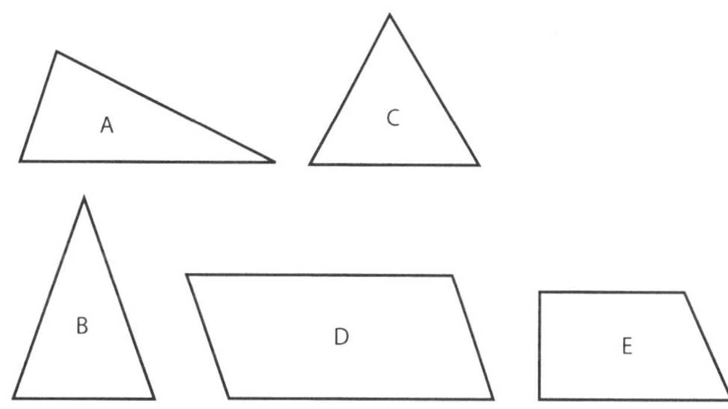

UNIT 31

Number and Algebra

SET 1 Basic

1 25 ÷ 5

2 6 × 3

3 ☐ tens + ☐ ones = 63

4 300 + 40 + 5

5 12c + 20c

6 4 cm = ☐ mm

7 48 ÷ 8

8 $1.95 = ☐ c

9 $11 × 10

10 $21 – $19

11 400 + 60 + 6

12 4000 m = ☐ km

13 How far is one quarter of 20 km?

14 54 ÷ 9

15 

How many 5 kg bags of sugar can be filled from an 80 kg drum?

☐ bags

SET 2 Rounding money

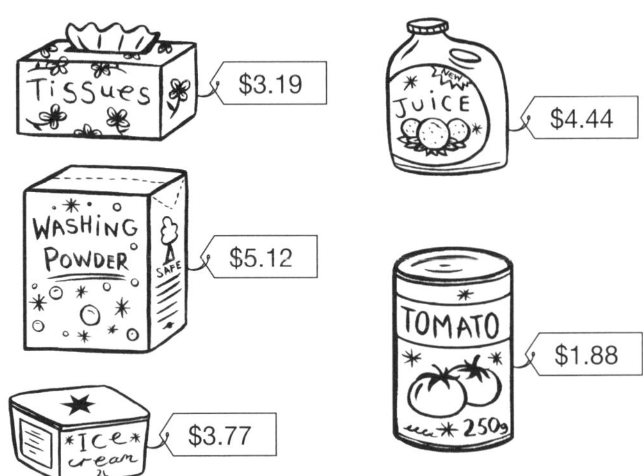

Complete the grid to calculate the cost of:

		Total price	Rounded price
1	Ice cream and washing powder	$	$
2	Tomatoes and tissues	$	$
3	Juice and washing powder	$	$
4	Washing powder and tomatoes	$	$
5	Ice cream and tissues	$	$
6	Tomatoes and juice	$	$
7	Ice cream and juice	$	$
8	Tissues and juice	$	$

Statistics and Probability Data investigation

Mr Peters' class collected data about age and long jump ability.

Name	Age	Long jump
Josh	10	2.1 m
Con	10.5	2.15 m
Janice	11	1.9 m
Toula	11.6	2.5 m
Jeremy	11.8	2.4 m
Sally	12.1	2.1 m
Ophra	12.6	2.35 m

Study the table then answer the questions.

1 Did the oldest person jump the longest?

2 Did the youngest person jump the shortest?

3 List some things that may affect a person's long jump ability.

____

____

____

Number and Algebra

SET 3 Algorithms for finding multiples

A number is divisible by 6 if the last digit of the number is even and the sum of its digits is a multiple of 3.

	Number	Is the last digit even?	Is the sum of the digits a multiple of 3?	Is the number a multiple of 6?
1	126			
2	183			
3	216			
4	248			

A number is divisible by 9 if the sum of all the digits in the number is a multiple of 9.

	Number	Is the sum of the digits a multiple of 9?	Is the number a multiple of 9?
5	128		
6	169		
7	783		
8	441		

SET 4 Extension

1 How much are 7 pencils at 2 for 30c?

2 740×3

3 What is the value of 6 in 136.75?

4 14 500 + 300 + 20

5 1.25 kg at $2 per kilogram

6 $6.72 – $2.48

7 How much is 75 cm of tape at $2 per metre?

8 $(7^2 - 25) + (\frac{3}{5}$ of 80)

9 If 6 pencils cost $1.20, how many could I buy for $4?

10 $\frac{7}{10} + \frac{2}{5}$

11 If the perimeter of a rectangle is 84 cm and one side is 24 cm long, what is the length of the sides that are not 24 cm?

12 How many fifths in 7.2?

13 Jacqui got 60% of her spelling test correst. If there were 100 words, how many did she get right?

14 What is the product of (4×10^2) and 2?

15 What is the difference between 10^2 and 4^2?

16 How many prime numbers are there between 20 and 40?

Space Estimating angles

Measure the size of each angle using a protractor.

1

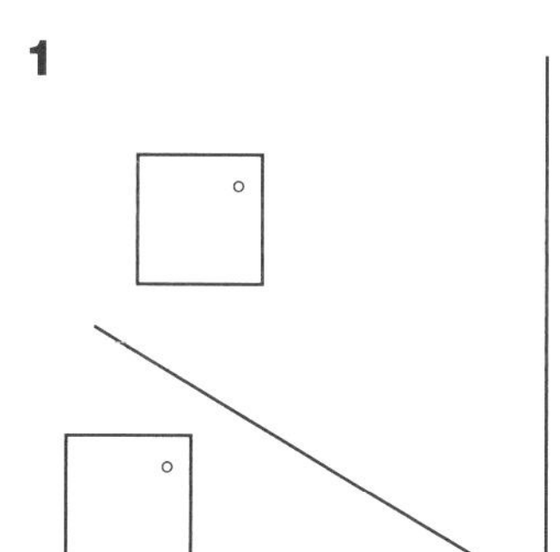

2

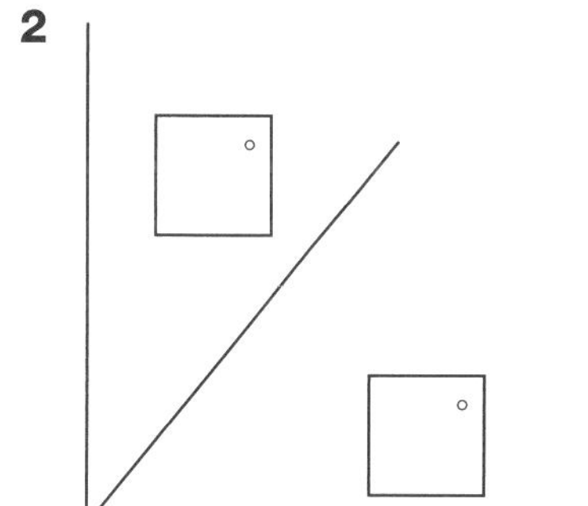

3

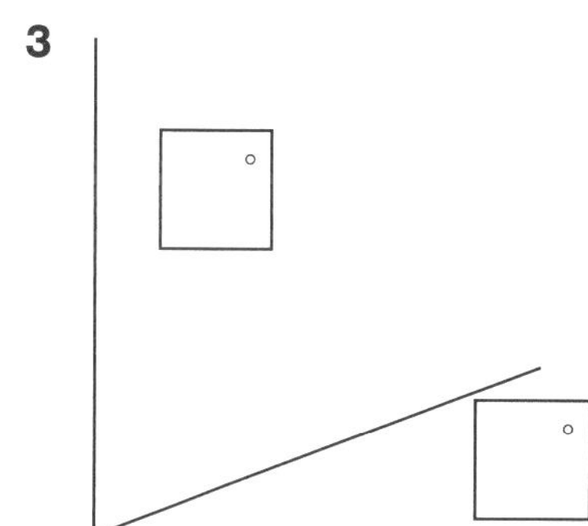

UNIT 32

Number and Algebra

SET 1 Basic

1 700 ☐ 350 = 350

2 8 × 4

3 27 − 14

4 48 ÷ 6

5 17 + 18

6 15 ☐ 3 = 5

7 15 × 3

8 47 − 8

9 42 ÷ 7

10 36 301, 36 307, 36 313, ☐

11 What is the product of 7 and 3?

12 What is the sum of 47 and 26?

13 How much is half of 46?

14 Triple 6.

15 

Oranges cost $6 a dozen. How much do 36 oranges cost?

$ ☐

SET 2 Decimals, percentages and fractions

	Fraction	Decimal	%
1	$\frac{77}{100}$		
2	$\frac{28}{100}$		
3	$\frac{7}{10}$		
4	$\frac{7}{100}$		
5	$\frac{1}{2}$		
6	$\frac{1}{4}$		
7	$\frac{3}{4}$		

Mathematical Reasoning

Order from smallest to largest.

8	$\frac{27}{100}$	30%	0.29	
9	35%	$\frac{53}{100}$	0.33	
10	$\frac{99}{100}$	9%	0.9	
11	0.54	$\frac{1}{2}$	49%	
12	4%	$\frac{3}{10}$	0.21	
13	$\frac{9}{100}$	90%	0.95	

Space Coordinates

1 Name the landmark at these coordinate points.

(C,6) ____________ (G,2) ____________ (Q,7) ____________ (T,3) ____________

2 Name any pair of coordinate points for Deep Lake ____________ and Dark Forest ____________.

3 Use the scale on the map to calculate these distances:

a Mountains to Forestville ____________ km. b Mountains to Hilltop ____________ km.

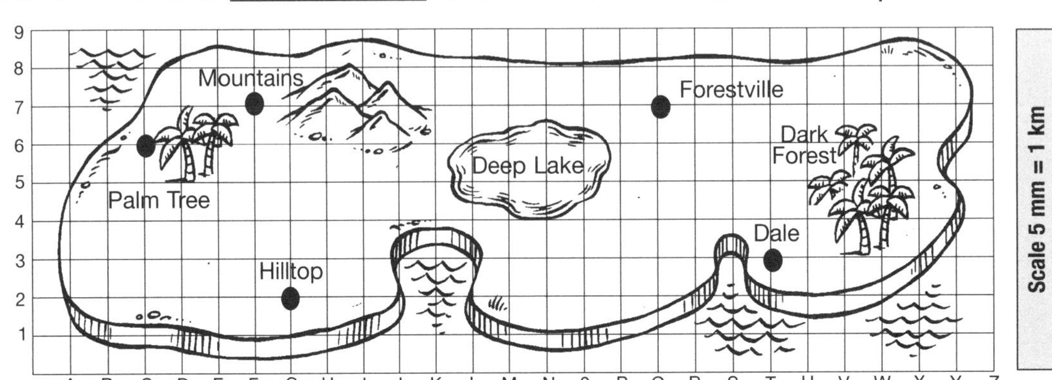

Number and Algebra

SET 3 Multiplication and division

Solve these multiplications using any strategy you wish.

1 210 × 4 = _______

2 140 × 3 = _______

3 220 × 2 = _______

4 330 × 4 = _______

5 225 × 5 = _______

6 244 × 4 = _______

7 258 × 3 = _______

8 318 × 3 = _______

Solve these divisions using any strategy you wish.

9 180 ÷ 5 = _______

10 248 ÷ 4 = _______

11 300 ÷ 6 = _______

12 321 ÷ 3 = _______

13 648 ÷ 8 = _______

14 618 ÷ 3 = _______

15 585 ÷ 5 = _______

16 981 ÷ 9 = _______

SET 4 Extension

1 How long is 4 decades minus 31 years?

2 What fraction of 1 minute is 45 seconds?

3 What fraction of 200 is 40?

4 Share 143 lollies between 6 people.

5 What is the perimeter of a rectangle with length 36 cm and width 23 cm?

6 $(\frac{3}{4} \times 32) \times (\frac{1}{5}$ of $40)$

7 Perimeter of a pentagon with 25 cm sides

8 How much is 4.25 kg of plaster at $16 per kilogram?

9 How much are 9 books at $4.05 each?

10 Which is the smallest: 10%, 0.1 or 0.09?

11 How many millilitres in 8.9 litres?

12 How many degrees in 2 triangles?

13 Write $27\frac{3}{100}$ as a decimal.

14 What is the area of a rectangle with length of 40 cm and a width of 7 cm?

15 $8.75 = 8\frac{3}{4}$ True or false?

16 Add all prime numbers between 6 and 20.

Measurement Recognising metric units

Solve the problems.

1 The length of a boat was 2 decametres and 2 metres long. How many metres is the boat in total? _______________

2 The length of a highway was 10 km and 642 m. How many metres in total was the highway?

____

3 The length of a piece of string was 8 m and 26 mm. How many millimetres was the piece of string? _______________

4 The total length of the hockey and football fields was 1 km and 31 m. What was the total length in metres? _______________

UNIT 33

Number and Algebra

SET 1 Basic

1 18 + 23

2 32 – 12

3 18 ☐ 2 = 36

4 18 ÷ 4

5 8 × 9

6 32 ☐ 36 = 68

7 45 ÷ 5

8 7 × 9

9 Write LXXIII in Hindu-Arabic numerals.

10 What is the difference between 43 and 15?

11 2 L = ☐ mL

12 How many tenths in $1\frac{1}{2}$?

13 How many months in spring and summer?

14 Write the set of factors for 12.

15

SET 2 Subtracting fractions from wholes

Subtract the fractions from the wholes.

1 $1 - \frac{1}{2} =$

2 $1 - \frac{1}{3} =$

3 $1 - \frac{1}{4} =$

4 $1 - \frac{1}{5} =$

5 $1 - \frac{1}{6} =$

6 $1 - \frac{1}{8} =$

7 $1 - \frac{1}{12} =$

8 $2 - \frac{1}{2} =$

9 $2 - \frac{1}{3} =$

10 $2 - \frac{1}{4} =$

11 $3 - \frac{1}{2} =$

12 $3 - \frac{1}{3} =$

13 $3 - \frac{1}{4} =$

14 $4 - \frac{1}{2} =$

Add or subtract the fractions.

15 $\frac{5}{8} + \frac{2}{8} =$

16 $\frac{3}{8} + \frac{4}{8} =$

17 $\frac{7}{12} + \frac{11}{12} =$

18 $\frac{11}{12} - \frac{7}{12} =$

19 $\frac{7}{8} - \frac{5}{8} =$

20 $\frac{4}{5} - \frac{3}{5} =$

21 Jim had a block of chocolate but gave away $\frac{3}{5}$ of it. How much did he have left? ☐

Space Sorting shapes

1 Colour the polygons.

2 Colour the quadrilaterals.

Number and Algebra

SET 3 Applying rules

Complete the bottom row by applying the rule.

1 ■ × 5 + 3 = ▲

■	1	2	3	4	5	6	7
▲							

2 ■ ÷ 2 + 10 = ▲

■	14	16	18	20	22	24	26
▲							

3 (■ + 12) ÷ 3 = ▲

■	6	9	12	15	18	21	24
▲							

4 (■ – 4) × 2 = ▲

■	11	12	13	14	15	16	17
▲							

5 Write your own rule and record your sequences in the grid.

Rule: ______________________________

■							
▲							

Mathematical Reasoning

SET 4 Extension

1 2.7 L = ☐ mL

2 Are 27 and 63 multiples of 9?

3 7.25 m = ☐ cm

4 $\frac{6}{8} + \frac{3}{4}$

5 (307 × 8) – 4

6 Write 23 $\frac{7}{10}$ as a decimal.

7 How much is 5.25 kg of meat at $6 per kilogram?

8 $(\frac{3}{4} \times 72) \times (\frac{3}{8} \times 24)$

9 If 9 books cost $63, how much would 27 cost?

10 Round 43.99 to the nearest whole number.

11 List the factors of 48.

12 What is the area of a rectangle with sides 14 cm and 5 cm?

13 $(\frac{3}{5} \times 150) - (\frac{3}{4} \times 44)$

14 How many 600 mL bottles are needed to fill a 3 L container?

15 Average 174, 326 and 250.

16 One quarter of 60 trees are eucalypts. How many are not?

Measurement Area units

Shade the box to show the best unit for measuring the area of the following.

	Item	cm^2	m^2	hectares
a	The area of a farm			
b	The area of a book cover			
c	The area of a large table			
d	The area of a city block			
e	The area of a car parking space			

UNIT 34

Number and Algebra

SET 1 Basic

1 54 ÷ 9

2 6 × 5

3 77 = ☐ tens + ☐ ones

4 400 + 60 + 5

5 48c + 6c

6 6 cm = ☐ mm

7 49 ÷ 7

8 $22.72 = ☐ c

9 $65 + $2.30

10 18 × 0

11 1600 + 40 + 8

12 One quarter of 28

13 72 ÷ 8

14 5c × 100

15

There are 44 bricks in a square metre. How many in a 10 m^2 wall?

☐ bricks

SET 2 Calculator division

Solve these divisions using a calculator.

1 $8\overline{)3608}$

2 $9\overline{)4059}$

3 $7\overline{)462}$

4 $8\overline{)364}$

5 $8\overline{)4348}$

6 $8\overline{)521}$

7 $4\overline{)3582}$

8 $4\overline{)4813}$

9 $6\overline{)4665}$

10 $3\overline{)2473}$

11 $9\overline{)3877}$

12 $6\overline{)1480}$

13 $15\overline{)390}$

14 $17\overline{)595}$

15 $26\overline{)936}$

Change these fractions to decimals.

16 $\frac{2}{5} = 0.$

17 $\frac{2}{8} = 0.$

18 $\frac{1}{4} = 0.$

19 $\frac{7}{10} = 0.$

20 $\frac{9}{12} = 0.$

21 $\frac{48}{100} = 0.$

Statistics and Probability Line graphs/tables

Use the data in the table to create a line graph to show a dog's mass over 7 years.

Age	0	1	2	3	4	5	6	7
Kgs	4	8	12	13	14	16	18	20

By how much did the dog's mass increase:

1 between 0 and 2 years?

2 between 2 and 5 years?

3 between 2 and 7 years?

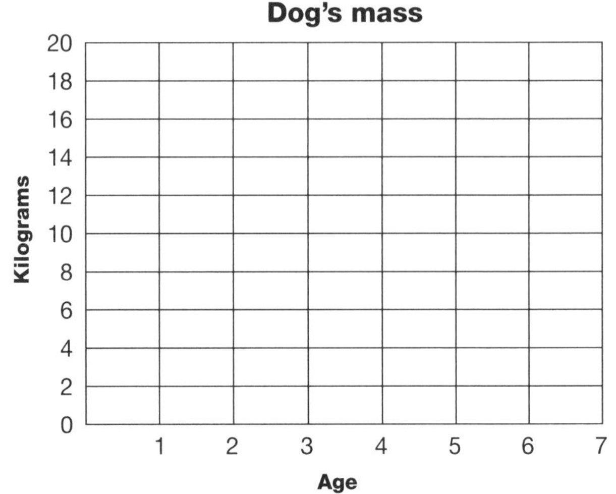

Number and Algebra

SET 3 Multiplying and dividing decimals

Complete this multiplication grid.

		× 10	× 100	× 1000
1	0.33			
2	1.87			
3	3.55			
4	2.62			

Complete this division grid.

		÷ 10	÷ 100	÷ 1000
5	145			
6	456			
7	905			
8	264			

Calculate the cost of this shopping bill.

	Shopping list	Cost
9	3 kg of grapes @ \$1.15/kg	
10	2 kg of grapes @ \$1.05/kg	
11	3 kg of grapes @ \$2.35/kg	
12	4 kg of grapes @ \$0.95/kg	
13	Total	

Mathematical Reasoning

SET 4 Extension

1 What is the value of 7 in 78.65?

2 Find the perimeter of a triangle with all sides 8 cm.

3 Share \$960 among 4 people.

4 Round 14.84 to the nearest tenth.

5 $180 - (\frac{7}{10} \text{ of } 90)$

6 Minutes between 10:05 am and 11:49 am

7 25% of \$900

8 How much is $8\frac{1}{2}$ m of timber at \$1.90 m?

9 How much is 0.8 m of ribbon at \$2 a metre?

10 What is the difference between 471 and 85?

11 How much are 7 trees at \$9.75 each?

12 Round 14 870 to the nearest 1000.

13 $(\frac{1}{3} \text{ of } 21) \times (\frac{7}{10} \text{ of } 120)$

14 How many tricycles will 381 wheels go on?

15 What is the value of 6 in 79.68?

16 $50 \times 40 + 10^2$

Space Properties of three-dimensional objects

Colour the description and its matching object the same colour.

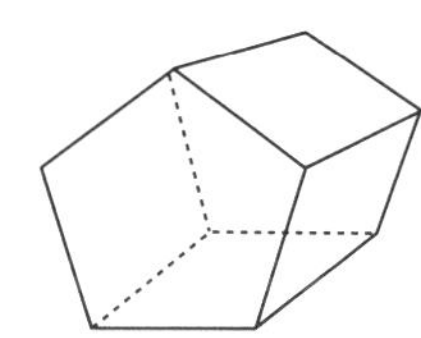

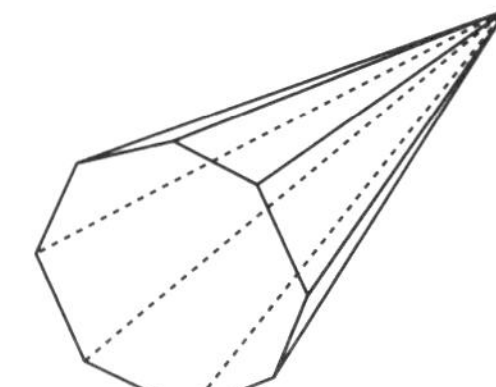

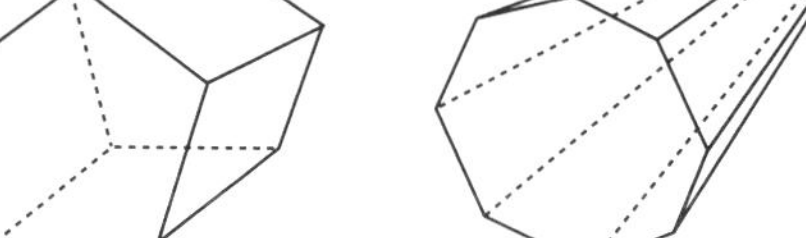

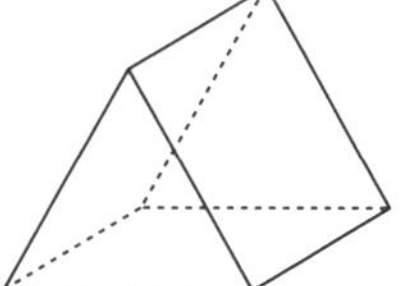

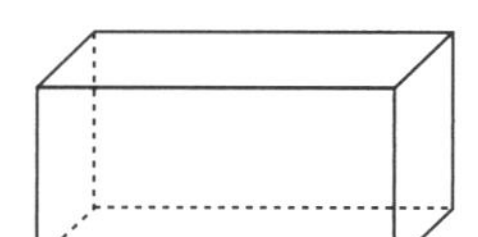

I am constructed from 2 pentagons and 5 rectangles. I am a prism.

I am constructed from 4 rectangles and 2 squares. I am a prism.

I am constructed from 2 equilateral triangles and 3 rectangles. I am a prism.

I am constructed from an octagon and 8 isosceles triangles. I am a pyramid.

Number and Algebra

SET 1 Basic

1 $35 \div 5$

2 $63 + 26$

3 $37 - 9$

4 $43 + 27$

5 $87 - 36$

6 9×4

7 8×7

8 $(3 + 6) \times 9$

9 What is the value of 7 in 37 306?

10 17 ☐ 3 = 51

11 87 + = ☐ 100

12 How many minutes in $3\frac{1}{2}$ hours?

13 Divide 50 by 4.

14 97 006 = ☐ + ☐ + ☐

15

How many 100 mL bottles are needed to fill a 2 L jug? ☐

SET 2 Improper fractions

Use the shapes to add the fractions.

1

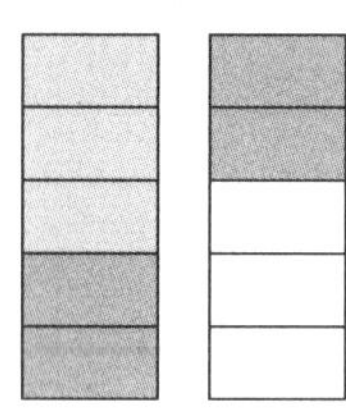

$\frac{3}{5} + \frac{4}{5} = \frac{\square}{5}$

2

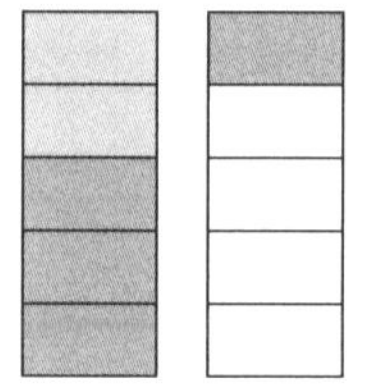

$\frac{2}{5} + \frac{4}{5} = \frac{\square}{5}$

Add these fractions. Record your answers as improper fractions and then mixed numerals.

3 $\frac{5}{10} + \frac{8}{10} = \frac{13}{10} = 1\frac{3}{10}$

4 $\frac{6}{8} + \frac{3}{8} = \square = \square$

5 $\frac{5}{8} + \frac{5}{8} = \square = \square$

6 $\frac{5}{8} + \frac{4}{8} = \square = \square$

7 $\frac{4}{10} + \frac{7}{10} = \square = \square$

8 $\frac{8}{10} + \frac{3}{10} = \square = \square$

9 $\frac{2}{4} + \frac{3}{4} = \square = \square$

10 $\frac{3}{6} + \frac{4}{6} = \square = \square$

11 $\frac{7}{10} + \frac{7}{10} = \square = \square$

Number and Algebra Sequences of fractions/decimals

Continue these sequences.

1	$\frac{1}{2}$	1	$1\frac{1}{2}$	2				
2	$\frac{1}{2}$	$1\frac{1}{4}$	2	$2\frac{3}{4}$				
3	$\frac{1}{5}$	$\frac{4}{5}$	$1\frac{2}{5}$	2				
4	$\frac{4}{10}$	$\frac{6}{10}$	$\frac{8}{10}$	1				

5	0.3	0.6	0.9	1.2				
6	0.8	1.3	1.8	2.3				
7	1	1.4	1.8	2.2				
8	2.5	2.9	3.	3.				

Number and Algebra

SET 3 Multiplication strategies

34 × 23
Think! 34 × 20 = 680
34 × 3 = 102
That's 782.

Use this strategy to complete these questions.

1 35 × 20 + 35 × 4

2 54 × 30 + 54 × 6

3 42 × 20 + 42 × 5

4 27 × 50 + 27 × 9

5 52 × 30 + 52 × 7

6 38 × 40 + 38 × 1

7 61 × 30 + 61 × 2

8 49 × 60 + 49 × 3

9 23 × 80 + 23 × 8

10 55 × 40 + 55 × 5

11 How much for 34 tickets at $29 each? 

12 How long is the fire hose if it consists of 23 lengths each 18 metres long? 

SET 4 Extension

1 What is the value of 7 in 39.76?

2 Round 12.71 to the nearest tenth.

3 How much is 9.5 mL of medicine at $12 a mL?

4 I had $1000 but spent $173. How much have I left?

5 Perimeter of a decagon with 9 cm sides

6 Write 134 $\frac{7}{100}$ as a decimal.

7 Difference between 13.76 m and 12.09 m

8 How many minutes from 11:05 am to 1:39 pm?

9 ($\frac{3}{4}$ of 12) × ($\frac{3}{5}$ of 70)

10 How many sides on 9 heptagons?

11 $\frac{6}{10} + \frac{1}{2} + \frac{3}{10}$

12 250 + ($\frac{3}{5}$ of 60)

13 17 cm and 6 mm = ☐ mm

14 Mr Cool planted 7 rows of 8 trees on Saturday and 16 rows of 4 trees on Sunday. How many trees will survive if one in six die? 

Measurement Comparing decimal measurements

Write these measurements in ascending order.

1 2.851 m, 3.612 m, 2.675 m, 1.891 m

____

2 0.861 mm, 0.768 mm, 0.852 mm

____

Write these measurements in descending order.

3 10.183 cm, 9.691 cm, 12.692 cm

____

4 0.128 mm, 0.361 mm, 0.113 mm

____

Maths helpers

Length

10 millimetres (mm) = 1 centimetre (cm)

100 centimetres (cm) = 1 metre (m)

1000 metres (m) = 1 kilometre (km)

Mass

1000 grams (g) = 1 kilogram (kg)

1000 kg = 1 tonne (t)

Capacity

1000 millilitres (mL) = 1 litre (L)

Time

60 seconds = 1 minute

60 minutes = 1 hour

24 hours = 1 day

7 days = 1 week

14 days = 1 fortnight

12 months = 1 year

52 weeks = 1 year

365 days = 1 year

366 days = 1 leap year

10 years = 1 decade

100 years = 1 century

Months of the year

Thirty days has September, April, June and November. All the rest have thirty-one, except February alone, which has twenty-eight days clear and twenty-nine days each leap year.

Seasons

Summer: December, January, February

Autumn: March, April, May

Winter: June, July, August

Spring: September, October, November

Roman numerals

1 = I

2 = II

3 = III

4 = IV

5 = V

6 = VI

7 = VII

8 = VIII

9 = IX

10 = X

20 = XX

30 = XXX

40 = XL

50 = L

60 = LX

70 = LXX

80 = LXXX

90 = XC

100 = C

500 = D

1000 = M

Multiplication facts

×	0	1	2	3	4	5	6	7	8	9	10
0	0	0	0	0	0	0	0	0	0	0	0
1	0	1	2	3	4	5	6	7	8	9	10
2	0	2	4	6	8	10	12	14	16	18	20
3	0	3	6	9	12	15	18	21	24	27	30
4	0	4	8	12	16	20	24	28	32	36	40
5	0	5	10	15	20	25	30	35	40	45	50
6	0	6	12	18	24	30	36	42	48	54	60
7	0	7	14	21	28	35	42	49	56	63	70
8	0	8	16	24	32	40	48	56	64	72	80
9	0	9	18	27	36	45	54	63	72	81	90
10	0	10	20	30	40	50	60	70	80	90	100

Addition facts

+	2	3	4	5	6	7	8	9	10	11	12
2	4	5	6	7	8	9	10	11	12	13	14
3	5	6	7	8	9	10	11	12	13	14	15
4	6	7	8	9	10	11	12	13	14	15	16
5	7	8	9	10	11	12	13	14	15	16	17
6	8	9	10	11	12	13	14	15	16	17	18
7	9	10	11	12	13	14	15	16	17	18	19
8	10	11	12	13	14	15	16	17	18	19	20
9	11	12	13	14	15	16	17	18	19	20	21
10	12	13	14	15	16	17	18	19	20	21	22
11	13	14	15	16	17	18	19	20	21	22	23
12	14	15	16	17	18	19	20	21	22	23	24

UNIT 1 Number and Algebra

SET 1

1 12
2 6
3 15
4 6
5 16
6 25
7 36
8 $13.68
9 3
10 4
11 23
12 40
13 6
14 7000
15 $51

SET 2

1 388
2 522
3 579
4 362
5 898
6 499
7 677
8 215
9 948
10 472
11 800, 814
12 200, 195
13 1500, 1492
14 1100, 1103
15 200, 207

SET 3

1 18
2 27
3 24
4 36
5 18 balls
6 36
7 54
8 300
9 27
10 400
11 yes
12 yes
13 90
14

Name	Rick	Sue	Gillian
Age	32	8	24

SET 4

1 300
2 37
3 2496
4 Yes
5 $9
6 86 421
7 5
8 $14
9 97 min
10 16
11 Half
12 1, 2, 4, 7, 14, 28
13 $4.50
14 $45
15 Twenty-six thousand

Space

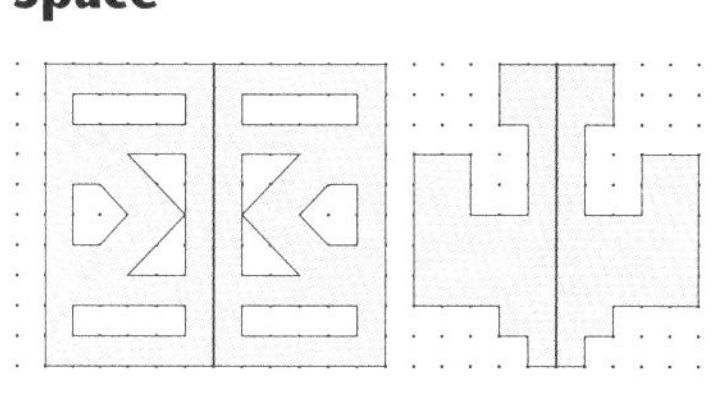

Measurement

1 km
2 mm
3 km
4 m
5 km
6 m

UNIT 2 Number and Algebra

SET 1

1 10
2 18
3 30
4 24
5 32
6 24
7 21
8 3215
9 2
10 3
11 55
12 40
13 10
14 6000
15 $35

SET 2

1 5839
2 9465
3 8633
4 9994
5 9063
6 8722
7 1285
8 3680
9 900
10 1800
11 5400
12 6700

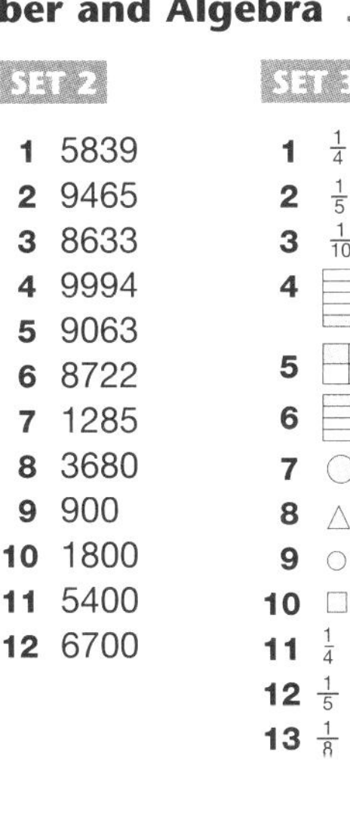

SET 3

1 $\frac{1}{4}$
2 $\frac{1}{5}$
3 $\frac{1}{10}$

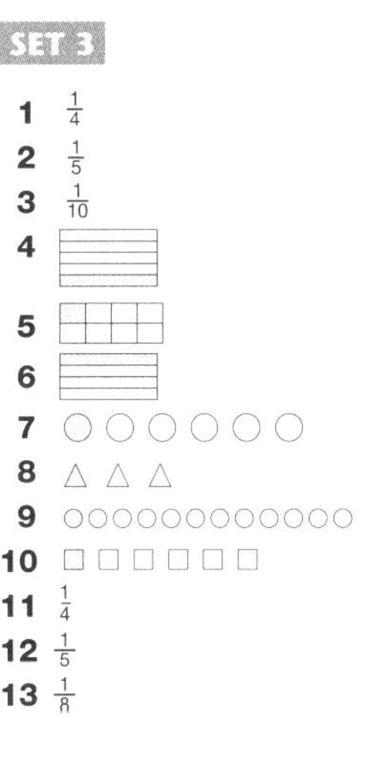

4
5
6
7 ○○○○○○
8 △△△
9 ○○○○○○○○○○○○○○○○
10 □□□□□□
11 $\frac{1}{4}$
12 $\frac{1}{5}$
13 $\frac{1}{8}$

SET 4

1 18
2 90
3 $49
4 39 168
5 6
6 10 h
7 6
8 307, 332
9 $\frac{1}{2}$
10 14 679
11 16
12 $21.60
13 1, 2, 4, 8, 16, 32
14 27
15 Thirty-nine thousand and six

Measurement

	Length	Width	Area (L × W)
1	4 cm	2 cm	8 cm^2
2	3 cm	2 cm	6 cm^2
3	3 cm	4 cm	12 cm^2
4	7 cm	3 cm	21 cm^2

Number and Algebra

1 4
2 8
3 3
4 5
5 5
6 4
7 3
8 4
9 5
10 6
11 6
12 5
13 8
14 7
15 9

UNIT 3 Number and Algebra

SET 1

1 11
2 13
3 11
4 40
5 1
6 54
7 14
8 421
9 6
10 7
11 12
12 12
13 300
14 3
15 9

SET 2

1 1561
2 3568
3 2080
4 5415
5 762
6 420
7 304
8 $6.45
9 308
10 $352

SET 3

1

2	4	6	8	10

2

3	6	9	12	15

3

4	8	12	16	20

4 True
5 False
6 True
7 True
8 False
9 False
10 True
11 True

SET 4

1 280
2 $288
3 395 cm
4 25
5 145
6 474
7 189
8 $36.48
9 $1.55
10 80
11 104
12 $12.65
13 $27
14 5599
15 370 cm
16

18	9	36	27	54	72	45	63
2	1	4	3	6	8	5	7

Space

Translate Rotate Reflect

Measurement

1 24 cm^3 Colour number 3
2 24 cm^3
3 21 cm^3
4 24 cm^3

Answers

UNIT 4 Number and Algebra

SET 1

1 11
2 17
3 14
4 3
5 14
6 24
7 24
8 468
9 6
10 10
11 28
12 14
13 400
14 7
15 78

SET 2

1 120 + 50 = 170
2 150 + 40 = 190
3 180 + 40 = 220
4 250 + 50 = 300
5 400 + 400 = 800
6 500 + 400 = 900
7 700 + 300 = 1000
8 1100 + 300 = 1400
9 2000 + 6000 = 8000
10 7000 + 3000 = 10 000
11 4000 + 3000 = 7000
12 10 000 + 4000 = 14 000
13 6000 + 5000 = 11 000

SET 3

1 Yes
2 No
3 No
4 Yes
5 Yes
6 Yes
7 1, $2\frac{1}{2}$, 4

SET 4

1 72
2 $6
3 $525
4 0.7
5 2700
6 240 min
7 115
8 39
9 $47.50
10 56
11 119
12 $12.25
13 30
14 53
15 65
16 21, 34, 55, 89

Space

1 a I have 6 sides of equal length, my opposite sides are parallel. I am a hexagon.
b I have 3 sides of equal length. I am an equilateral triangle.

2

Measurement

Hands on

UNIT 5 Number and Algebra

SET 1

1 13
2 13
3 1
4 3
5 40
6 54
7 32
8 671
9 6
10 7
11 56
12 19
13 5
14 30
15 7

SET 2

1 16, 160
2 24, 240
3 32, 320
4 40, 400
5 36, 360
6 28, 280
7 24, 240
8 36, 360
9 48, 480
10 60, 600
11 54, 540
12 42, 420
13 $300
14 $280
15 $270
16 $220

SET 3

1 797 km
2 1424 km
3 552 km
4 75 km
5 1349 km
6 627 km
7 No: $36 is the correct change.

SET 4

1 2654
2 1350
3 97 631
4 5846
5 92
6 16th
7 $2.15
8 18
9 2023
10 $1.80
11 No
12 24
13 $39
14 103
15 46 398
16 Acute

Space

Right Acute Reflex Obtuse Straight

Statistics and Probability

1 Unlikely
2 Equal chance
3 Impossible
4 Likely
5 Certain

UNIT 6 Number and Algebra

SET 1

1 12
2 13
3 5
4 2
5 24
6 40
7 27
8 680
9 7
10 8
11 36
12 20
13 7
14 4 ones
15 $2.50

SET 2

1
3 × 5 = 15
5 × 3 = 15
15 ÷ 3 = 5
15 ÷ 5 = 3

2
4 × 6 = 24
6 × 4 = 24
24 ÷ 6 = 4
24 ÷ 4 = 6

3
6 × 8 = 48
8 × 6 = 48
48 ÷ 8 = 6
48 ÷ 6 = 8

SET 3

1 Rule: add 4
23, 27, 31, 35, 39, 43, 47
2 Rule: add 5
2, 7, 12, 17, 22, 27, 32
3 Rule: subtract 3
59, 56, 53, 50, 47, 44, 41
4 Rule: double
1, 2, 4, 8, 16, 32, 64
5 Rule: add 6

Squares	1	2	3	4	5	6	7
Sticks	6	12	18	24	30	36	42

SET 4

1 About 1250
2 470
3 8500
4 16
5 42 668
6 25th
7 $1.90
8 72
9 126
10 $21
11 7 tenths + 3 hundredths
12 $48
13 $19.35
14 9000
15 56 969
16 5 to 9 or 8:55

Number and Algebra

1 18 ÷ 3 = 6

2 25 ÷ 5 = 5

Measurement

	Digital	24-hour
1	10:20 am	1020
2	11:15 am	1115
3	3:25 pm	1525
4	9:50 pm	2150
5	10:45 pm	2245

UNIT 7 Number and Algebra

SET 1

1 42
2 40
3 12
4 38
5 27
6 21
7 5
8 6
9 80
10 35
11 5
12 60 min
13 80c
14 00 000
15 10

SET 2

1 972
2 1808
3 1182
4 518
5 1092
6 2480
7 2009
8 3045

SET 3

1 $\frac{2}{4}$
2 $\frac{3}{6}$
3 $\frac{6}{12}$ $\frac{1}{4}$
4 $\frac{2}{8}$
5 $\frac{4}{16}$
6 $\frac{1}{2}$ $\frac{2}{4}$
7 $\frac{1}{3}$ $\frac{2}{6}$
8 $\frac{1}{2}$ $\frac{5}{10}$
9 $\frac{2}{4}$ $\frac{5}{10}$
10 $\frac{1}{5}$ $\frac{2}{10}$
11 $\frac{2}{8}$ $\frac{1}{4}$

SET 4

1 0.81
2 19
3 19
4 80
5 20
6 2356, 4127, 26 007
7 40
8 Yes
9 $17.50
10 32
11 1425
12 1, 2, 4, 5, 8, 10, 20, 40
13 7
14 2.5
15 20 girls, 12 boys

Statistics and Probability

1–4 Hands on

Measurement

1 2000 m
2 km
2 4000 m
4 km

UNIT 8 Number and Algebra

SET 1

1 5
2 8
3 21
4 33
5 27
6 16
7 60
8 28
9 30
10 2607
11 35
12 63
13 4
14 15
15 12

SET 2

1 2222
2 1132
3 2113
4 2112
5 2221
6 $3974
7 $2724
8 $1214

SET 3

1 64, 67, 70, 73, 76, 79, 82, 85
2 355, 360, 365, 370, 375, 380 385, 390
3 6, 14, 22, 30, 38, 46, 54, 62
4 21, 20, 19, 18, 17, 16, 15, 14
5 9, 7, 5, 3

SET 4

1 30
2 $45
3 True
4 1.0
5 4
6 22
7 165 mm
8 23 467
9 Rectangles (includes squares)
10 2809
11 No
12 4 am
13 $2
14 7:25 am
15 Thirty thousand and twenty-seven

Statistics and Probability

1 125 cm
2 140 cm
3 130 cm
4 15 cm
5 No

Space

1 north west
2 north east
3 south east
4 south west
5 700 m
6 500 m
7 600 m

UNIT 9 Number and Algebra

SET 1

1 24
2 7
3 48
4 8
5 40
6 5
7 18
8 8
9 11
10 24
11 60
12 17
13 4
14 2
15 12

SET 2

1 48
2 48
3 70
4 72
5 170
6 90
7 700
8 130
9 3.2, 6.4
10 8.4, 16.8
11 5, 10
12 62.4, 124.8
13 10.4, 20.8, 41.6
14 64.8, 129.6, 259.2
15 28.6, 57.2, 114.4
16 62.4, 124.8, 249.6

SET 3

1 5
2 4
3 40
4 9
5 5
6 10
7 60
8 100
9 200
10 1000
11 Will 48, Susan 36
Laura 24, Tom 12, Kate 6

SET 4

1 About 700
2 13.8
3 750
4 $22.50
5 999 m
6 471
7 $360
8 14
9 1392, 1422
10 $18.50
11 $18.35
12 $377
13 7905
14 Anna: $30
Bree: $15
Cath: $10
Deni: $5

Space

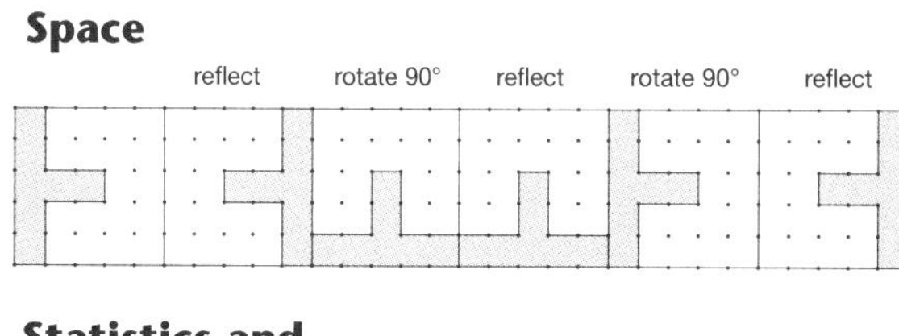

Statistics and Probability

1 A 2 A 3 B 4 C
5 B and C

Answers

UNIT 10 Number and Algebra

SET 1

1 21
2 18
3 54
4 6
5 ÷
6 ÷
7 +
8 108
9 –
10 28
11 120 min
12 15c
13 50
14 1
15 96

SET 2

1 42 064
2 62 613
3 59 312
4 88 147
5 93 235
6 79 918
7 57 779
8 69 194
9 $45 110

SET 3

1 True
2 True
3 True
4 True
5 True
6 False
7 True
8 True
9 False
10 False
11 3, 6, 9, 12, 15, 18
12 4, 8, 12, 16, 20, 24
13 6, 12, 18, 24, 30, 36
14 7, 14, 21, 28, 35, 42
15 9, 18, 27, 36, 45, 54

SET 4

1 False
2 29
3 9
4 1, 3, 5, 9, 15, 45
5 8
6 7
7 2700
8 No
9 $16
10 14
11 27
12 35 427, 28 963, 27 301
13 $\frac{3}{4}$
14 $2.25
15 Twenty thousand two hundred and six

Measurement

1 24 cm^2
2 12 cm^2
3 20 cm^2

Space

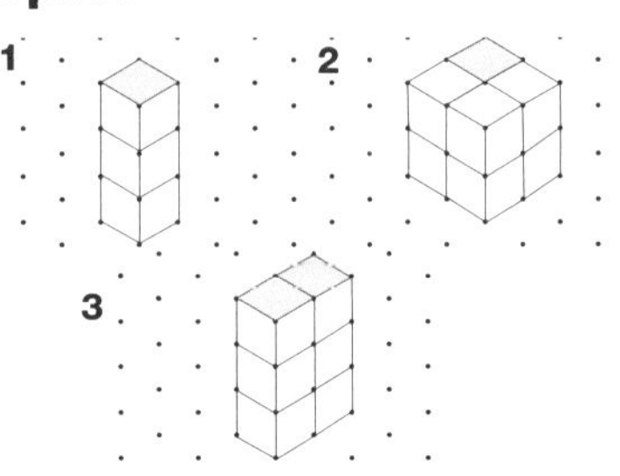

UNIT 11 Number and Algebra

SET 1

1 22
2 +
3 15
4 –
5 35
6 ÷
7 8
8 ×
9 210
10 60
11 63
12 16
13 56c
14 100 yrs
15 210

SET 2

1 $7 \times 3 = 21$ ✓
2 $7 \times 7 = 49$ ✓
3 $6 \times 9 = 54$ ✓
4 $8 \times 7 = 54$ ✗
5 $8 \times 4 = 36$ ✗
6 $9 \times 9 = 81$ ✓
7 $6 \times 8 = 56$ ✗
8 $5 \times 9 = 45$ ✓
9 $6 \times 7 = 42$ ✓
10 24
11 16
12 11 r 2
13 27
14 18
15 16 r 3
16 26
17 11
18 16 r 2

SET 3

1 $30 \times 3 = 90$
2 $50 \times 5 = 250$
3 $100 \times 2 = 200$
4 $30 \times 5 = 150$
5 $40 \times 3 = 120$
6 $40 \times 5 = 200$
7 $20 \times 7 = 140$
8 $30 \times 9 = 270$
9 $50 \times 6 = 300$
10 $30 \times 7 = 210$
11 $40 \times 8 = 320$
12 $40 \times 6 = 240$
13 $50 \times 9 = 450$
14 $40 \times 5 = 200$
15 $40 \times 8 = 320$

SET 4

1 425
2 Yes
3 6
4 147
5 41
6 $11
7 296
8 $16
9 5
10 7496, 20 206, 53 224
11 42 cm
12 42 hundreds
13 150 km

Statistics and Probability

1 1
2 7
3 24
4 Game 4

Space

Colour 1 and 5 the same
Colour 2 and 7 the same
Colour 3 and 8 the same
Colour 4 and 6 the same

UNIT 12 Number and Algebra

SET 1

1 44
2 $12
3 6 tens 8 ones
4 $17
5 25
6 239
7 $7.70
8 49
9 $2.60
10 4000
11 9
12 30
13 $40
14 100
15 $7\frac{1}{2}$

SET 2

1 $18 700
2 $26 850
3 $16 095
4 $10 360
5 $10 755
6 $26 455

SET 3

(1-6 any four of the following answers)
1 1, 2, 4, 5, 10, 20
2 1, 2, 4, 8, 16
3 1, 2, 3, 4, 6, 12
4 1, 2, 4, 5, 8, 10, 20, 40
5 1, 2, 4, 8, 16, 32
6 1, 2, 3, 5, 6, 10, 15, 30
7 Yes
8 Yes
9 No
10 Yes
11 1, 2, 3, 4, 6, 8, 12, 24

SET 4

1 56
2 $5.03
3 262
4 1431
5 68 cm
6 $15.22
7 7000
8 $3.60
9 $1.60
10 14, 21, 28, 35, 42, 49
11 $42.60
12 75
13 155
14 860
15 152 hundreds
16 20
17 43 km

Space

Shape	Translate	Rotate	Reflect
N	N	Z	И

Measurement

1 30 cm^3
2 72 cm^3

UNIT 13 Number and Algebra

SET 1

1 12
2 90
3 40
4 4
5 7
6 21
7 24
8 608
9 90
10 23
11 $7.70
12 10
13 621
14 $30
15 56

SET 2

1 815
2 843
3 2964
4 1770
5 3059
6 2478
7 1680
8 200 km
9 $1400

SET 3

1 15
2 18
3 19
4 24
5 30
6 31
7 20
8 15
9 9
10 26
11 T
12 F (48)
13 T
14 T
15 F (53)
16 T
17 F (9)
18 F (90)
19 F (300)
20 F (323)
21 Any pair that has a sum of 12 e.g., 12 + 0, 11 + 1, 10 + 2, 8 + 4

SET 4

1 25 000
2 1829
3 1, 2, 3, 6, 7, 14, 21, 42
4 146
5 154
6 $39
7 7 tenths
8 38 000
9 10 × 9
10 4658
11 $2.50
12 7:10 pm
13 C cm
14 Discussion
$3.95 × 10 = $39.50

Space

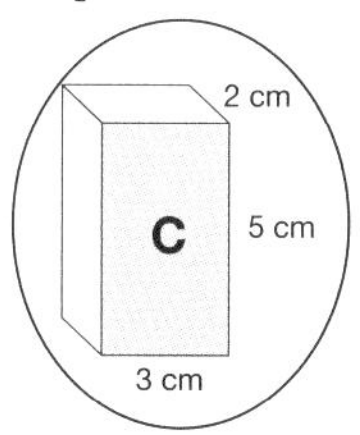

Measurement

1 0930
2 1200
3 1330
4 1600
5 1900
6 A Country Practice
7 Night Court

UNIT 14 Number and Algebra

SET 1

1 40
2 ÷
3 ×
4 445
5 $3.60
6 45
7 4
8 77
9 85
10 24
11 107
12 14
13 200
14 45
15 5

SET 2

1 60, 20, 80, 40, 100, 50, 90
2 20, 10
3 12, 6
4 32, 16
5 100, 50
6 44, 22
7 80, 40
8 28, 14, 7
9 40, 20, 10
10 100, 50, 25
11 80, 40, 20
12 48, 24, 12

SET 3

1 $\frac{6}{5}$
2 $\frac{12}{10}$
3 $\frac{14}{8}$
4 $\frac{9}{5}$
5 $\frac{17}{10}$
6 $\frac{11}{8}$
7 $\frac{9}{6}$

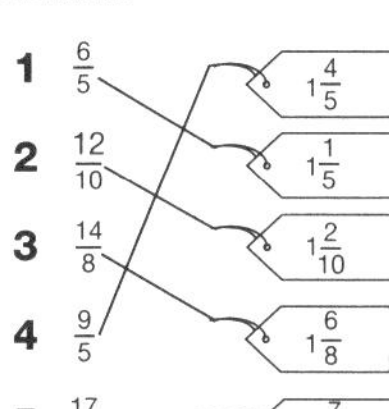

8 $1\frac{3}{8}$
9 $1\frac{2}{5}$
10 $1\frac{0}{10}$
11 $1\frac{2}{6}$
12 $1\frac{2}{8}$
13 $1\frac{6}{12}$

SET 4

1 19 000
2 99
3 $650
4 1, 2, 4, 8, 16, 32
5 $9\frac{1}{2}$ hrs
6 $2.55
7 40
8 80%
9 80c
10 6 cm
11 15
12 150
13 90c each
14 8
15 24
16 A square with sides of 4 m

Statistics and Probability

1 Red
2 Pink
3 No
4 No
5 Red

Measurement

1 18 cm
2 24 cm
3 26 cm

UNIT 15 Number and Algebra

SET 1

1 1305
2 21
3 +
4 27
5 –
6 54
7 6
8 ×
9 ÷
10 56
11 1000 mL
12 25c
13 7000
14 10 yrs
15 161

SET 2

1 708
2 728
3 1470
4 3552
5 1296
6 4494
7 1460
8 5936
9 1446
10 3320 g
11 Emma: $8
Mark: $48

SET 3

1 1, 2, 4, 8, 16
2 1, 2, 4, 5, 10, 20

SET 4

1 60
2 21 000
3 10 800
4 8500
5 27 226
6 6720
7 23, 29
8 2.5
9 215 min
10 $2\frac{1}{3}$
11 9 hundredths
12 20 cm
13 20
14 40
15 $3.50
16 13 km

Statistics and Probability

Possible solution

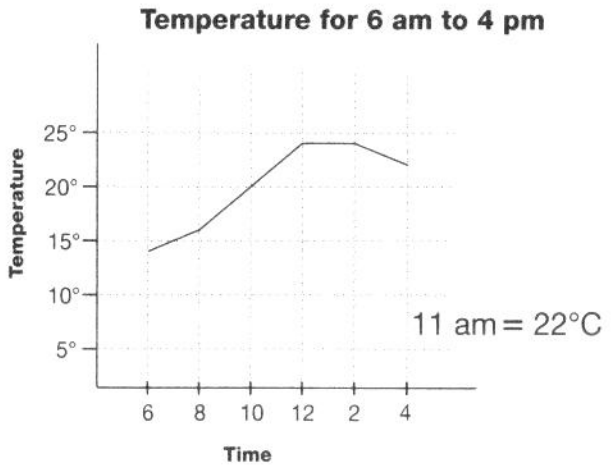

Measurement

1 8 m^2
2 16 m^2
3 28 m^2

Answers

UNIT 16 Number and Algebra

SET 1

1 12, 16
2 63
3 48
4 4
5 5
6 21
7 30
8 673
9 170
10 60
11 10
12 196
13 2000
14 $3.60
15 $6.60

SET 2

1 432
2 122
3 160 r 4
4 221 r 2
5 230
6 210
7 324
8 114
9 131
10 242
11 254
12 124
13 $189
14 $77

SET 3

	Visual	Tenth	Hundredth	Decimal
1		$\frac{9}{10}$	$\frac{90}{100}$	0.9
2		$\frac{7}{10}$	$\frac{70}{100}$	0.7
3		$\frac{3}{10}$	$\frac{30}{100}$	0.3

4 $\frac{93}{100}$
5 0.06
6 $\frac{70}{100}$
7 $\frac{21}{100}$
8 0.99
9 0.87
10 0.7
11 0.1

SET 4

1 6 r 4
2 $51 800
3 $26.98
4 25 cm^2
5 246
6 $70.24
7 425
8 40
9 Same
10 0.07
11 23.3
12 $1.05
13 9 cm
14 37 942
15 23 632 or 32 623

Space

1 B, G
2 C
3 D, E
4 F
5 A

Measurement

	Less than 1 t	About 1 t	More than 1 t
elephant			✓
25 children		✓	
television	✓		
car		✓	
bulldozer			✓
refrigerator	✓		

UNIT 17 Number and Algebra

SET 1

1 5
2 4
3 +
4 12
5 38
6 14
7 ×
8 16
9 4
10 785
11 127
12 31
13 1
14 $4.20
15 Various responses

SET 2

1 0.16, 0.21, 0.23
2 0.17, 0.52, 0.73
3 0.16, 0.79, 0.97
4 0.01, 0.17, 0.27
5 0.02, 0.03, 0.04
6 1.07, 1.27, 2.71
7 2.53, 3.25, 5.23
8 1.27, 2.12, 7.21
9 1.67, 1.76, 3.16
10 0.26, 0.28
11 1.26, 1.28
12 4.08, 4.1

SET 3

1 1, 2, 5, 10
2 1, 2, 3, 4, 6, 12
3 1, 3, 5, 15
4 1, 2, 4, 8, 16
5 1, 2, 4, 5, 10, 20
6 1, 2, 3, 4, 6, 8, 12, 24

(Possible combinations)

7 23 × 12 = 23 × 4 × 3 = 276
8 31 × 15 = 31 × 5 × 3 = 465
9 41 × 12 = 41 × 6 × 2 = 492
10 52 × 15 = 52 × 5 × 3 = 780
11 43 × 14 = 43 × 7 × 2 = 602
12 39 × 12 = 39 × 6 × 2 = 468
13 40 × 20 = 40 × 5 × 4 = 800
14 32 × 16 = 32 × 4 × 4 = 512
15 44 × 16 = 32 × 8 × 2 = 704
16 33 × 20 = 33 × 5 × 4 = 660

SET 4

1 516
2 33
3 230
4 2900
5 48 602, 37 100, 29 501
6 28 cm^2
7 52
8 150
9 $9
10 632
11 20 000
12 16 r 1
13 3.30, 3.36
14 45
15 250 000
16 28 cm

Statistics and Probability

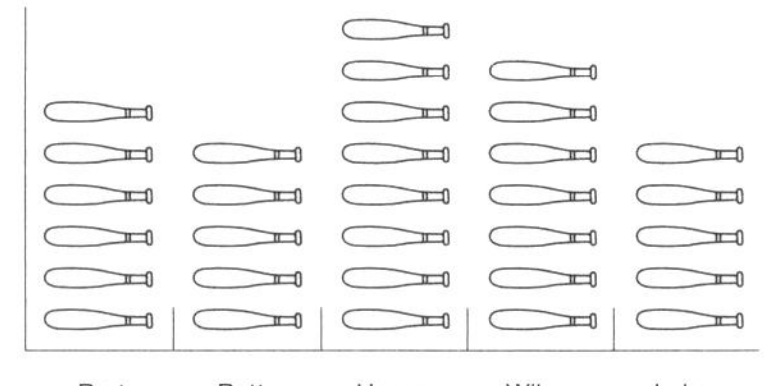

Space

Possible descriptions:

1 Hexagonal prism. 8 faces, 12 vertices, 18 edges.
2 Pentagonal pyramid. 6 faces, 6 vertices, 10 edges.

UNIT 18 Number and Algebra

SET 1

1 54
2 6
3 17
4 5
5 ÷
6 +
7 –
8 ×
9 231
10 3
11 30 000
12 75c
13 8
14 54
15 3

SET 2

1 True
2 True
3 False
4 True
5 False
6 True
7 True
8 True
9 False
10 True
11 False
12 True

SET 3

1 $\frac{1}{4}, \frac{1}{2}, \frac{3}{4}, 1, 1\frac{1}{4}, 1\frac{1}{2}, 1\frac{3}{4}, 2$
2 $\frac{1}{3}, \frac{2}{3}, 1, 1\frac{1}{3}, 1\frac{2}{3}, 2$
3 $\frac{1}{5}, \frac{2}{5}, \frac{3}{5}, \frac{4}{5}, 1, 1\frac{1}{5}, 1\frac{2}{5}$
4 $\frac{1}{8}, \frac{2}{8}, \frac{3}{8}, \frac{4}{8}, \frac{5}{8}, \frac{6}{8}, \frac{7}{8}, 1$
5 $\frac{1}{10}, \frac{2}{10}, \frac{3}{10}, \frac{4}{10}, \frac{5}{10}, \frac{6}{10}, \frac{7}{10}, \frac{8}{10}, \frac{9}{10}, 1$
6 $\frac{1}{2}, 1, 1\frac{1}{2}, 2, 2\frac{1}{2}$

SET 4

1 29
2 79 000
3 7, 11, 13, 17, 19
4 4.17
5 800 g
6 134 min
7 37 200
8 162 cm
9 15
10 55
11 $\frac{5}{10}, \frac{50}{100}, \frac{1}{2}$
12 $1.35
13 135 sec
14 35 256, 1479, 256
15 2000 g or 2 kg

Number and Algebra

	A	B	C	D	
1	Date	Item	Cost	Balance	
2	1 June	Opening			$400
3	7 June	Food	$160	(D2 – C3)	$240
4	10 June	Petrol	$40	(D3 – C4)	$200
5	11 June	Bills	$120	(D4 – C5)	$80
6	20 June	Drinks	$30	(D5 – C6)	$50
7	29 June	Movies	$20	(D6 – C7)	$30

Statistics and Probability

1 3
2 4
3 2
4 3
5 Alphabetical order of surnames

UNIT 19 Number and Algebra

SET 1

1 25
2 14
3 48
4 5
5 18
6 19
7 415
8 195
9 24
10 22
11 17
12 63
13 21
14 28
15 $6

SET 2

1 <
2 <
3 <
4 <
5 >
6 >
7 <
8 =
9 >
10 >
11 9.514, 9.516
12 4.397, 4.399
13 6.798, 6.800
14 8.399, 8.401
15 3.600, 3.602

SET 3

Hands on

SET 4

1 $37.50
2 64 962
3 42
4 243
5 $10 \times 10 \times 10$
6 11 000
7 About 700
8 9601
9 15
10 28.9
11 750
12 8659
13 $27
14 3.5
15 46 398
16 $67

Number and Algebra

Tens	Ones	•	Tenths	Hundredths	Thousandths
	8	•	3	2	1
	7	•	8	3	5
1	6	•	1	9	7
1	8	•	6	1	4
	6	•	0	1	1

Space

(a) 50° (b) 40°
(c) 70° (d) 40°
(e) 30° (f) 110°

UNIT 20 Number and Algebra

SET 1

1 20
2 12
3 35
4 5
5 14
6 371
7 144
8 24
9 28
10 33
11 72
12 81
13 7
14 48
15 144

SET 2

1 124
2 162
3 117
4 136
5 120 r1
6 136 r2
7 93 r3
8 183 r2
9 113 r5
10 127 r2
11 107 r4
12 110 r1
13 122 r7
14 112 r1
15 174 r1
16 136 r1

SET 3

1 $9935
2 $196
3 $415
4 $2076
5 $208
6 35c

SET 4

1 5.9
2 34 985
3 40
4 $3.60
5 4000
6 10 000
7 $1.54
8 3856
9 10:45
10 10 006
11 80
12 7.23
13 6200
14 $133
15 57 409
16 19:25

Space

	Length	Width	Area
Carpet	8 m	6 m	48 m^2
Lino	4 m	3 m	12 m^2

Statistics and Probability

1 About $200
2 About $50
3 Yes
4 About $15

UNIT 21 Number and Algebra

SET 1

1 25
2 9
3 41
4 80
5 11
6 32
7 2 r 1
8 6000
9 +
10 –
11 38
12 48
13 XXVII
14 $1
15 22

SET 2

1 $1032
2 $1548
3 $1372
4 $1720
5 Yes
6 Sydney
7 $2654

SET 3

1 $\frac{2}{4}$
2 $\frac{3}{4}$
3 $\frac{2}{3}$
4 $\frac{3}{5}$
5 $\frac{4}{5}$
6 $\frac{9}{10}$
7 $\frac{9}{12}$
8 $\frac{54}{100}$
9 $1\frac{2}{5}$
10 $2\frac{2}{4}$
11 $3\frac{7}{10}$
12 $4\frac{2}{3}$
13 $5\frac{5}{8}$

SET 4

1 950 cm
2 4.07
3 4500 g
4 4.52
5 56 cm
6 $14
7 $16.50
8 8.32
9 9 m
10 No
11 Length: 48 m
Width: 48 m

Number and Algebra

	A	B	C	
1	Date	Deposit	Subtotal	
2	22 Mar	$217.30	(= C2)	$217.30
3	28 Mar	$133.40	(= C2 + B3)	**$350.70**
4	4 Apr	$168.10	(= C3 + B4)	**$518.80**
5	13 Apr	$321.70	(= C4 + B5)	**$840.50**
6	19 Apr	$298.90	(= C5 + B6)	**$1139.40**

Space

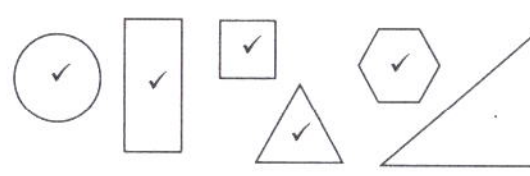

Answers

UNIT 22 Number and Algebra

SET 1

1 ×
2 +
3 10 000
4 41
5 4
6 24
7 54
8 72
9 35
10 6
11 13
12 October
13 $9
14 1224
15 144

SET 2

1 $15.38
2 $205.95
3 17 517
4 8747
5 28 199
6 1894
7 Blake and Greenfield

SET 3

1 $44
2 $99
3 $66
4 $220
5 $176

SET 4

1 Yes
2 14.09
3 80
4 3.8
5 5 cm
6 $54
7 176 hundredths
8 64 cm
9 0.34
10 860
11 $\frac{1}{4}$
12 47
13 $24
14 $25.90
15 60

Statistics and Probability

1 4 km
2 8 km
3 12 km
4 0
5 3 km per hour

Measurement

1 7:25 am
2 7:20 am
3 10:16 am
4 9:40 am
5 30 mins

UNIT 23 Number and Algebra

SET 1

1 50
2 48
3 18
4 7
5 54
6 47
7 29
8 2
9 0
10 21
11 29
12 11
13 9
14 23
15 $70

SET 2

1 25 000
2 63 000
3 36 000
4 64 000
5 18 000
6 18 000
7 49 000
8 24 000
9 40 000
10 56 000
11 27 402
12 31 152
13 25 053
14 58 770
15 $38 270

SET 3

1 $\frac{10}{100}$, 0.10, 10%
2 $\frac{25}{100}$, 0.25, 25%
3 $\frac{7}{10}$, 0.7, 70%
4 $\frac{20}{100}$, 0.2, 20%
5 $\frac{1}{2}$, 0.5, 50%
6 $\frac{1}{4}$, 0.25, 25%
7 $\frac{3}{4}$, 0.75, 75%
8 $\frac{27}{100}$, 0.29, 30%
9 0.33, 35%, $\frac{53}{100}$
10 9%, 0.9, $\frac{99}{100}$
11 49%, $\frac{1}{2}$, 0.54
12 4%, 021, $\frac{3}{10}$
13 $\frac{9}{100}$, 90%, 0.95
14 0.03, 7%, $\frac{70}{100}$

SET 4

1 $1\frac{2}{5}$
2 7000
3 2
4 5158
5 356
6 $22.40
7 45
8 $7.60
9 $18
10 11 pm
11 360°
12 18 800
13 184
14 48 cm
15 12 629
16 64 kg

Statistics and Probability

Hands on. One example below.

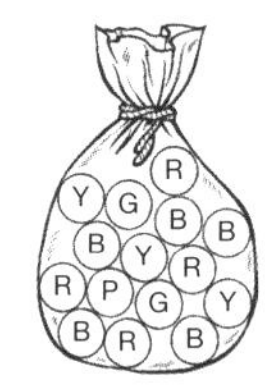

Number and Algebra

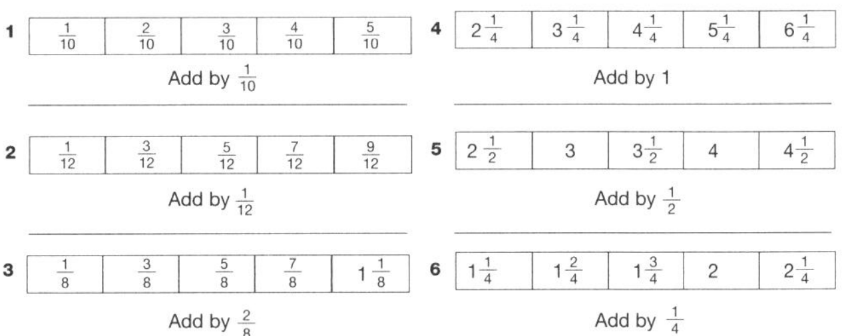

UNIT 24 Number and Algebra

SET 1

1 5
2 36
3 46
4 42
5 37
6 32
7 32
8 +
9 ×
10 60
11 41
12 54
13 XXXIX
14 13 765
15 $12.50

SET 2

1 70 ÷ 7 = 10
2 50 ÷ 10 = 5
3 90 ÷ 9 = 10
4 80 ÷ 10 = 8
5 150 ÷ 10 = 15
6 4
7 40
8 400
9 3
10 30
11 300
12 2
13 20
14 200
15 10
16 100
17 1000

SET 3

1 4
2 8
3 6
4 41
5 51
6 21
7 3
8 [10]
9 △7 + (12)
10 (12) + [10] = 22
11 [10] + (12) + ⬡3 = 25
12 ⬡3 × (12) + [10] = 46
13 ([10] − ⬡3) × (12) = 84

SET 4

1 925
2 48
3 $1.36
4 360°
5 False
6 48
7 119 cm
8 $29.50
9 90 cm
10 5852
11 $137.50
12 30 000
13 212 min
14 30
15 $72
16 2245

Space

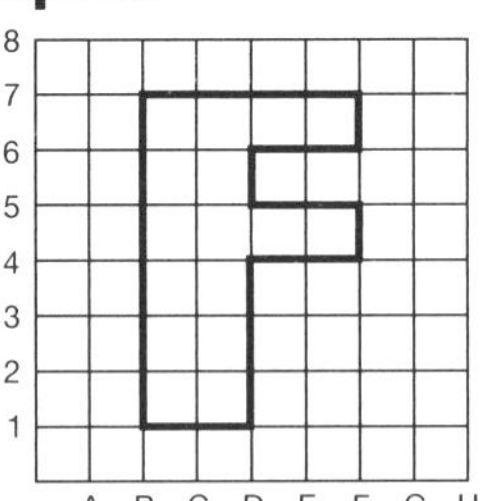

Measurement

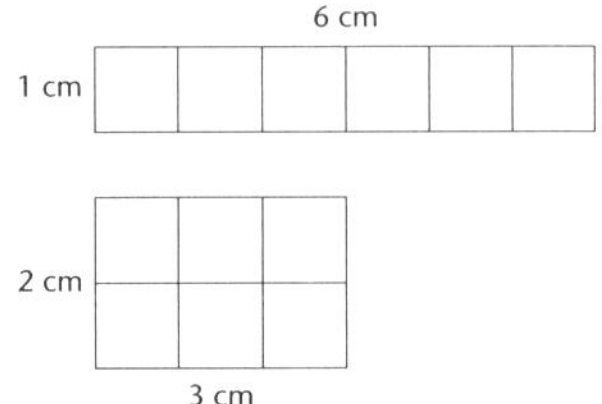

UNIT 25 Number and Algebra

SET 1

1 21
2 4
3 29
4 36
5 4
6 54
7 9
8 3000
9 ×
10 −
11 70
12 20
13 61
14 200
15 $210

SET 2

×	10	100	1000
1 2	20	200	2000
2 4	40	400	4000
3 8	80	800	8000
4 11	110	1100	11 000
5 12	120	1200	12 000

6 500
7 1020
8 1160
9 6150
10 8220
11 10 780
12 $9730

SET 3

1 2, $2\frac{1}{2}$
2 1, $1\frac{1}{4}$
3 $1\frac{1}{3}$, $1\frac{2}{3}$
4 $3\frac{1}{3}$, $4\frac{1}{3}$
5 3, $3\frac{1}{3}$
6 $1\frac{7}{8}$, $2\frac{1}{8}$
7 0.4, 0.45
8 0.6, 0.68
9 0.52, 0.47
10 0.2, 0.25

SET 4

1 30
2 27
3 True
4 20
5 180°
6 570 km
7 5988
8 3.7
9 87
10 85 kg
11 Baseball: 20
Soccer: 30
Hockey: 10

Measurement

	Perimeter	Area
A	10 cm	5 cm^2
B	12 cm	5 cm^2
C	16 cm	7 cm^2
D	16 cm	13 cm^2

Space

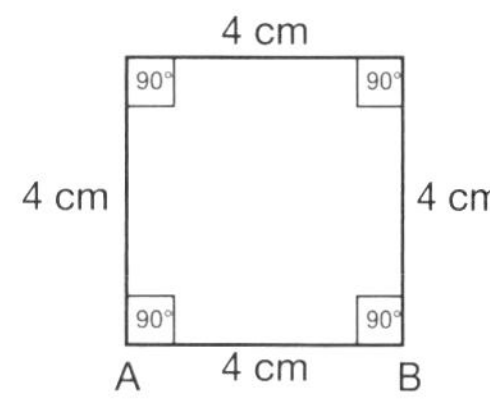

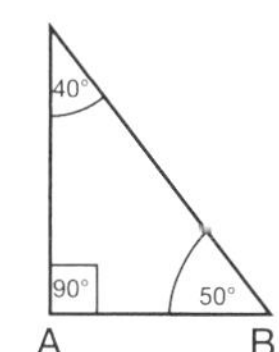

UNIT 26 Number and Algebra

SET 1

1 24
2 31
3 9
4 56
5 30
6 14
7 19
8 5
9 5
10 18
11 $4.40
12 4
13 5
14 20
15 148

SET 2

1 528
2 1643
3 1350
4 891
5 1722
6 1890
7 324
8 682
9 6138
10 1008

SET 3

1 37
2 5
3 36
4 4
5 8
6 1
7 45
8 3
9 =
10 ≠
11 =
12 ≠
13 =
14 ≠
15 =

SET 4

1 12 km
2 28 cm
3 427
4 $\frac{6}{100}$
5 16 300
6 Tuesday 2 September
7 0
8 $225
9 3750
10 150
11 $20.10
12 $144
13 $8.82

Statistics and Probability

1

2 90 cm
3 115 cm
4 5 years

Measurement

1 20
2 5
3 5
4 40

UNIT 27 Number and Algebra

SET 1

1 30
2 54
3 6
4 28
5 21
6 431
7 4000
8 $4
9 6
10 $2.40
11 180
12 13
13 36
14 3
15 $879

SET 2

1 120
2 103
3 208 r 2
4 138 r 4
5 105
6 302 r 1
7 105
8 104
9 87 r 4
10 100
11 78
12 96
13 34 m^2
14 105
15 $101

SET 3

1 $5.75
2 $1.20
3 $3.65
4 $0.85
5 $3404
6 $4248

SET 4

1 $24.50
2 9:20
3 17 700
4 60
5 5000
6 15.8
7 164
8 464
9 5
10 18
11 $14.40
12 45
13

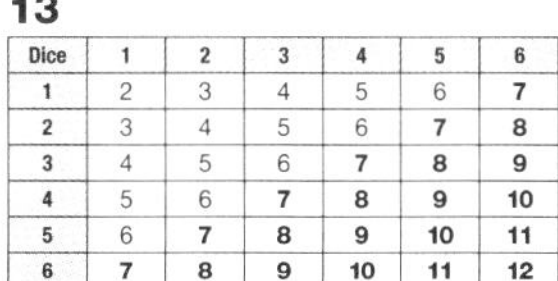

Dice	1	2	3	4	5	6
1	2	3	4	5	6	7
2	3	4	5	6	7	8
3	4	5	6	7	8	9
4	5	6	7	8	9	10
5	6	7	8	9	10	11
6	7	8	9	10	11	12

7 is the most common score.

Measurement

1 2000 mL
2 7000 mL
3 2500 mL
4 2250 mL
5 3250 mL
6 7100 mL
7 9300 mL

Measurement

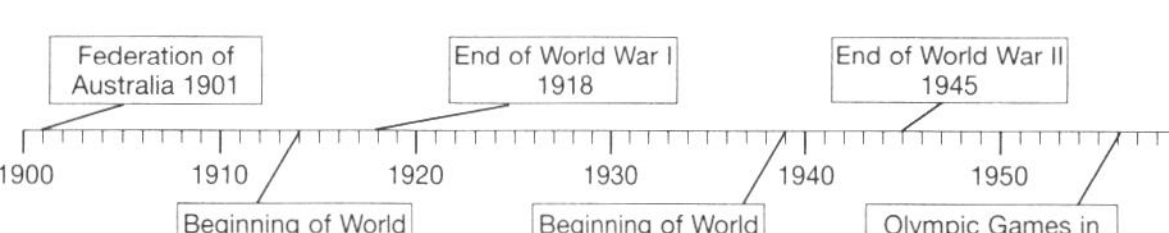

Answers

UNIT 28 Number and Algebra

SET 1

1 28
2 $2.90
3 12 m
4 1, 2, 4, 8, 16
5 64th
6 94
7 $120
8 3
9 0
10 $6.50
11 548
12 0
13 34
14 7 tens, 2 ones
15 360

SET 2

1 6
2 40
3 25
4 7
5 40
6 54
7 87
8 1000
9 9
10 38
11 $20
12 700 mL

SET 3

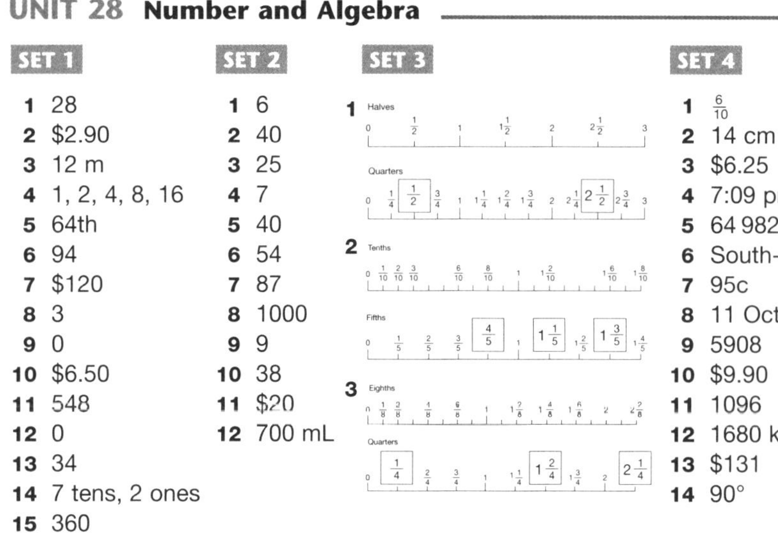

SET 4

1 $\frac{6}{10}$
2 14 cm
3 $6.25
4 7:09 pm
5 64 982
6 South-east
7 95c
8 11 October
9 5908
10 $9.90
11 1096
12 1680 km
13 $131
14 90°

Measurement

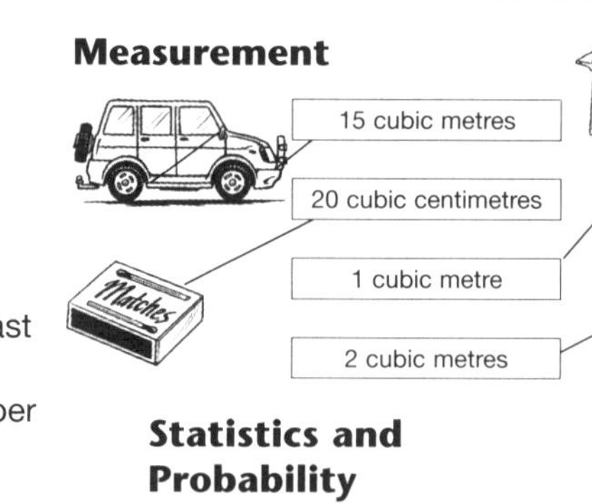

Statistics and Probability

1 20°C
2 17°C
3 10°C
4 12°C
5 About 21°C

UNIT 29 Number and Algebra

SET 1

1 5
2 50
3 8
4 72
5 27
6 74
7 32
8 5 r 1
9 600
10 ÷
11 +
12 $23
13 27
14 2 hrs
15 Thursday

SET 2

1 $\frac{9}{10}$
2 $\frac{3}{10}$
3 $\frac{5}{10}$
4 $\frac{4}{8}$
5 $\frac{5}{8}$
6 $\frac{3}{5}$
7 $\frac{4}{5}$
8 $\frac{2}{5}$
9 $\frac{5}{6}$
10 $\frac{2}{12}$
11 $\frac{5}{6}$
12 $\frac{11}{12}$
13 $\frac{4}{6}$
14 $\frac{7}{8}$
15 $\frac{6}{18}$
16 $\frac{12}{18}$ $\frac{6}{18}$

SET 3

1 10
2 30
3 40
4 40
5 10
6 8
7 7
8 70
9 10
10 25
11 Shade 25 circled red
12 Shade 15 circles blue
13 Shade 20 circles yellow
14 Shade 40 circles green

SET 4

1 27.43
2 $49
3 $57
4 39 cm
5 10
6 13.7
7 30
8 $\frac{1}{8}$
9 $42.50
10 20
11 4 metres

Number and Algebra

1	1089	4000	5000	10 000
2	2189	8000	10 000	20 000
3	3588	16 000	20 000	40 000
4	4768	20 000	25 000	50 000
5	5217	20 000	25 000	50 000
6	6891	28 000	35 000	70 000

7	397	400	4	100	99 r 1
8	789	800	8	100	98 r 5
9	791	800	4	200	197 r 3
10	149	150	5	30	29 r 4
11	254	250	5	50	50 r 4

Measurement

1 7 days
2 Fort Denison
3 eight
4 1.81 m
5 Thursday 7th Jan.

UNIT 30 Number and Algebra

SET 1

1 57
2 31
3 40
4 8
5 56
6 7
7 8
8 ÷
9 +
10 35
11 6
12 23
13 13 771
14 ninety-five
15 18 kg

SET 2

1 $\frac{5}{8}$
2 $\frac{5}{8}$
3 1
4 $\frac{3}{4}$
5 $\frac{5}{8}$
6 $\frac{3}{8}$
7 $\frac{7}{8}$
8 $\frac{5}{8}$
9 $\frac{7}{8}$
10 $1\frac{2}{8}$
11 $1\frac{1}{2}$
12 $1\frac{6}{8}$

SET 3

1 $24
2 $22
3 $20
4 It decreases
5 Mon $211
6 Tues $138
7 Wed $168
8 Thurs $161
9 Fri $169

SET 4

1 16 cm^2
2 5
3 500 m
4 1179
5 7 tens
6 $\frac{1}{4}$
7 37 400
8 170 cm
9 57
10 $11.60
11 17.17
12 $21.60
13 81
14 54 months
15 Straight
16 $45
17 $4800
18 $240

Space

1 A
2 B
3 C
4 D
5 E

Space

UNIT 31 Number and Algebra

SET 1

1 5
2 18
3 6 tens + 3 ones
4 345
5 32c
6 40 mm
7 6
8 195
9 $110
10 $2
11 466
12 4
13 5 km
14 6
15 16

SET 2

		Total price	Rounded price
1	Ice cream and washing powder	$ 8.89	$ 8.90
2	Tomatoes and tissues	$ 5.07	$ 5.05
3	Juice and washing powder	$ 9.56	$ 9.55
4	Washing powder and tomatoes	$ 7	$ 7
5	Ice cream and tissues	$ 6.96	$ 6.95
6	Tomatoes and juice	$ 6.32	$ 6.30
7	Ice cream and juice	$ 8.21	$ 8.20
8	Tissues and juice	$ 7.63	$ 7.65

SET 3

	Number	Is the last digit even?	Is the sum of the digits a multiple of 3?	Is the number a multiple of 6?
1	126	Yes	Yes	Yes
2	183	No	Yes	No
3	216	Yes	Yes	Yes
4	248	Yes	No	No

	Number	Is the sum of the digits a multiple of 9?	Is the number a multiple of 9?
5	128	No	No
6	169	No	No
7	783	Yes	Yes
8	441	Yes	Yes

SET 4

1 $1.05
2 2220
3 6 ones
4 14 820
5 $2.50
6 $4.24
7 $1.50
8 72
9 20
10 $1\frac{1}{10}$
11 18 cm
12 36
13 60
14 000
15 84
16 4 (23, 29, 31, 37)

Space

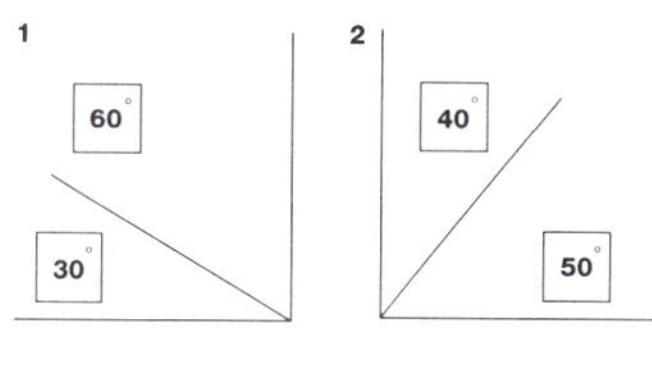

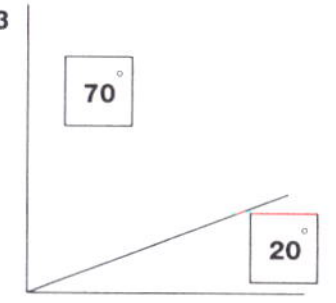

Statistics and Probability

1 No
2 No
3 Hands on

UNIT 32 Number and Algebra

SET 1

1 –
2 32
3 13
4 8
5 35
6 ÷
7 45
8 39
9 6
10 36 319
11 21
12 73
13 23
14 18
15 $10

SET 2

1 $\frac{77}{100}$, 0.77, 77%
2 $\frac{28}{100}$, 0.28, 28%
3 $\frac{7}{10}$, 0.7, 70%
4 $\frac{7}{100}$, 0.07, 7%
5 $\frac{1}{2}$, 0.5, 50%
6 $\frac{1}{4}$, 0.25, 25%
7 $\frac{3}{4}$, 0.75, 75%
8 $\frac{27}{100}$, 0.29, 30%
9 0.33, 35%, $\frac{53}{100}$
10 9%, 0.9, $\frac{99}{100}$
11 49%, $\frac{1}{2}$, 0.54
12 4%, 0.21, $\frac{3}{10}$
13 $\frac{9}{100}$, 90%, 0.95

SET 3

1 840
2 420
3 440
4 1320
5 1125
6 976
7 774
8 954
9 36
10 62
11 50
12 107
13 81
14 200
15 117
16 109

SET 4

1 9 yrs
2 $\frac{3}{4}$
3 $\frac{1}{5}$
4 23 r 5
5 118 cm
6 192
7 125 cm
8 $68
9 $36.45
10 0.09
11 8900 mL
12 360°
13 27.03
14 200 cm^2
15 True
16 67 (7 + 11 + 13 + 17 + 19)

Space

1 (C,6) = Palm Tree
(G,2) = Hilltop
(Q,7) = Forestville
(T,3) = Dale
2 Possible answers:
Deep Lake = (L,5), (L,6), (M,4), (M,5), (M,6), (N,4), (N,5), (N,6), (O,5), (O,6)
Dark Forest = (U,4), (U,5), (U,6), (V,4), (V,5), (V,6), (W,3), (W,4), (W,5), (W,6)
3 a 11 km b 5 km

Measurement

1 22 m
2 10 612 m
3 8026 mm
4 1031 m

UNIT 33 Number and Algebra

SET 1

1 41
2 20
3 ×
4 4 r 2
5 72
6 +
7 9
8 63
9 73
10 28
11 2000 mL
12 15
13 6
14 1, 2, 3, 4, 6, 12
15 48

SET 2

1 $\frac{1}{2}$
2 $\frac{2}{3}$
3 $\frac{3}{4}$
4 $\frac{4}{5}$
5 $\frac{5}{6}$
6 $\frac{7}{8}$
7 $\frac{11}{12}$
8 $1\frac{1}{2}$
9 $1\frac{2}{3}$
10 $1\frac{3}{4}$
11 $2\frac{1}{2}$
12 $2\frac{2}{3}$
13 $2\frac{3}{4}$
14 $3\frac{1}{2}$
15 $\frac{7}{8}$
16 $\frac{7}{8}$
17 $\frac{18}{12}$
18 $\frac{4}{12}$
19 $\frac{2}{8}$
20 $\frac{1}{5}$
21 $\frac{2}{5}$

SET 3

1 8, 13, 18, 23, 28, 33, 38
2 17, 18, 19, 20, 21, 22, 23
3 6, 7, 8, 9, 10, 11, 12
4 14, 16, 18, 20, 22, 24, 26
5 Hands on

SET 4

1 2700
2 Yes
3 725
4 $1\frac{2}{4}$ or $1\frac{1}{2}$
5 2452
6 23.7
7 $31.50
8 486
9 $189
10 44
11 1, 2, 3, 4, 6, 8, 12, 16, 24, 48
12 70 cm^2
13 57
14 5
15 250
16 45

Space

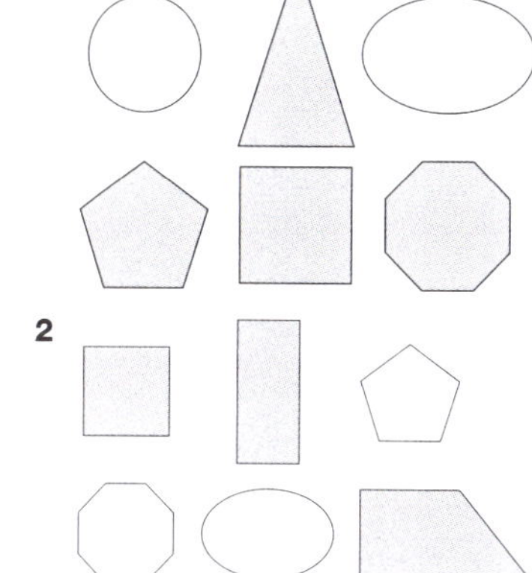

Measurement

Hands on

Answers

UNIT 34 Number and Algebra

SET 1

1 6
2 30
3 7 tens, 7 ones
4 465
5 54c
6 60
7 7
8 2272
9 $67.30
10 0
11 1648
12 7
13 9
14 $5
15 440

SET 2

1 451
2 451
3 66
4 45.5
5 543.5
6 65.125
7 895.5
8 1203.25
9 777.5
10 824.333
11 430.777
12 246.666
13 26
14 35
15 36
16 0.4
17 0.25
18 0.25
19 0.7
20 0.75
21 0.48

SET 3

		× 10	× 100	× 1000
1	0.33	3.3	33	330
2	1.87	18.7	187	1870
3	3.55	35.5	355	3550
4	2.62	26.2	262	2620

		÷ 10	÷ 100	÷ 1000
5	145	14.5	1.45	0.145
6	456	45.6	4.56	0.456
7	905	90.5	9.05	0.905
8	264	26.4	2.64	0.264

	Shopping list	Cost
9	3 kg of grapes @ $1.15/kg	$3.45
10	2 kg of grapes @ $1.05/kg	$2.10
11	3 kg of grapes @ $3.35/kg	$7.05
12	4 kg of grapes @ $0.95/kg	$3.80
13	Total	$16.40

SET 4

1 70
2 24 cm
3 $240
4 14.8
5 117
6 104 min
7 $225
8 $16.15
9 $1.60
10 386
11 $68.25
12 15 000
13 588
14 127
15 6 tenths
16 2100

Space

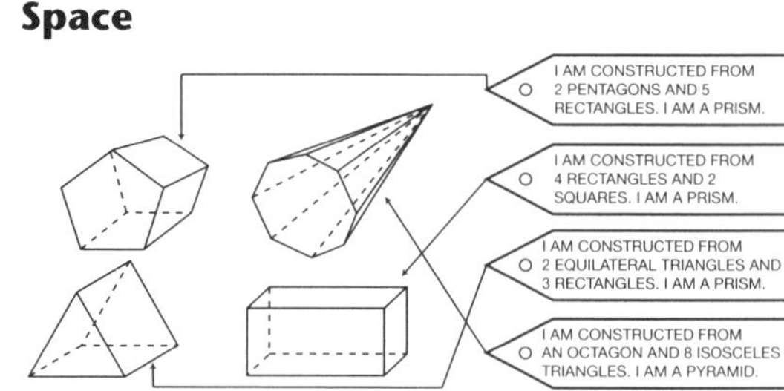

Statistics and Probability

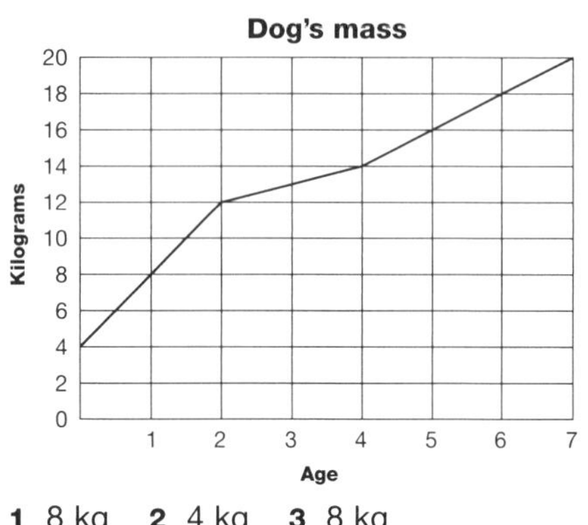

1 8 kg **2** 4 kg **3** 8 kg

UNIT 35 Number and Algebra

SET 1

1 7
2 89
3 28
4 70
5 51
6 36
7 56
8 81
9 7000
10 ×
11 13
12 210 min
13 12 r 2
14 90 000 + 7000 + 6
15 20

SET 2

1 $\frac{7}{5}$
2 $\frac{6}{5}$
3 $\frac{13}{10} = 1\frac{3}{10}$
4 $\frac{9}{8} = 1\frac{1}{8}$
5 $\frac{10}{8} = 1\frac{2}{8}$
6 $\frac{9}{8} = 1\frac{1}{8}$
7 $\frac{11}{10} = 1\frac{1}{10}$
8 $\frac{11}{10} = 1\frac{1}{10}$
9 $\frac{5}{4} = 1\frac{1}{4}$
10 $\frac{7}{6} = 1\frac{1}{6}$
11 $\frac{14}{10} = 1\frac{4}{10}$

SET 3

1 840
2 1944
3 1050
4 1593
5 1924
6 1558
7 1952
8 3087
9 2024
10 2475
11 $986
12 414 m

SET 4

1 $\frac{7}{10}$
2 12.7
3 $114
4 $827
5 90 cm
6 134.07
7 1.67 m
8 154 min
9 378
10 63
11 $1\frac{4}{10}$ or $1\frac{2}{5}$
12 286
13 176
14 100

Number and Algebra

1	$\frac{1}{2}$	1	$1\frac{1}{2}$	2	$2\frac{1}{2}$	3	$3\frac{1}{2}$	4
2	$\frac{1}{2}$	$1\frac{1}{4}$	2	$2\frac{3}{4}$	$3\frac{1}{2}$	$4\frac{1}{4}$	5	$5\frac{3}{4}$
3	$\frac{1}{5}$	$\frac{4}{5}$	$1\frac{2}{5}$	2	$2\frac{3}{5}$	$3\frac{1}{5}$	$3\frac{4}{5}$	$4\frac{2}{5}$
4	$\frac{4}{10}$	$\frac{6}{10}$	$\frac{8}{10}$	1	$1\frac{2}{10}$	$1\frac{4}{10}$	$1\frac{6}{10}$	$1\frac{8}{10}$
5	0.3	0.6	0.9	1.2	1.5	1.8	2.1	2.4
6	0.8	1.3	1.8	2.3	2.8	3.3	3.8	4.3
7	1	1.4	1.8	2.2	2.6	3	3.4	3.8
8	2.5	2.9	3.3	3.7	4.1	4.5	4.9	5.3

Measurement

1 1.891 m, 2.675 m, 2.851 m, 3.612 m
2 0.768 mm, 0.852 mm, 0.861 mm
3 12.692 cm, 10.183 cm, 9.691 cm
4 0.361 mm, 0.128 mm, 0.113 mm